MPEG-4
Beyond Conventional Video Coding
Object Coding, Resilience, and Scalability

Copyright © 2006 by Morgan & Claypool

All rights reserved. No part of this publication may be reproduced, stored in a retrieval system, or transmitted in any form or by any means—electronic, mechanical, photocopy, recording, or any other except for brief quotations in printed reviews, without the prior permission of the publisher.

MPEG-4 Beyond Conventional Video Coding: Object Coding, Resilience, and Scalability
Mihaela van der Schaar, Deepak S Turaga and Thomas Stockhammer
www.morganclaypool.com

1598290428   paper   van der Schaar/Turaga/Stockhammer
1598290436   ebook   van der Schaar/Turaga/Stockhammer

DOI 10.2200/S00011ED1V01Y200508IVM004

A Publication in the Morgan & Claypool Publishers' series
SYNTHESIS LECTURES ON IMAGE, VIDEO & MULTIMEDIA PROCESSING
Lecture #4
ISSN print: 1559-8136
ISSN online: 1559-8144

First Edition
10 9 8 7 6 5 4 3 2 1

Printed in the United States of America

MPEG-4 Beyond Conventional Video Coding
Object Coding, Resilience, and Scalability

Mihaela van der Schaar
University of California, Los Angeles

Deepak S Turaga
IBM T.J. Watson Research Center

Thomas Stockhammer
Munich University of Technology

SYNTHESIS LECTURES ON IMAGE, VIDEO & MULTIMEDIA PROCESSING #4

MORGAN & CLAYPOOL PUBLISHERS

ABSTRACT

An important merit of the MPEG-4 video standard is that it not only provided tools and algorithms for enhancing the compression efficiency of existing MPEG-2 and H.263 standards but also contributed key innovative solutions for new multimedia applications such as real-time video streaming to PCs and cell phones over Internet and wireless networks, interactive services, and multimedia access. Many of these solutions are currently used in practice or have been important stepping-stones for new standards and technologies. In this book, we do not aim at providing a complete reference for MPEG-4 video as many excellent references on the topic already exist. Instead, we focus on three topics that we believe formed key innovations of MPEG-4 video and that will continue to serve as an inspiration and basis for new, emerging standards, products, and technologies. The three topics highlighted in this book are object-based coding and scalability, Fine Granularity Scalability, and error resilience tools. This book is aimed at engineering students as well as professionals interested in learning about these MPEG-4 technologies for multimedia streaming and interaction. Finally, it is not aimed as a substitute or manual for the MPEG-4 standard, but rather as a tutorial focused on the principles and algorithms underlying it.

KEYWORDS
MPEG-4, object-coding, fine granular scalability, error resilience, robust transmission.

Contents

1. Introduction 1
2. Interactivity Support: Coding of Objects with Arbitrary Shapes 5
   2.1 Shape Coding 8
       2.1.1 Binary Shape Coding 8
       2.1.2 Grayscale Shape Coding 18
   2.2 Texture Coding 20
       2.2.1 Intracoding 20
       2.2.2 Intercoding 22
   2.3 Sprite Coding 24
   2.4 Encoding Considerations 27
       2.4.1 Shape Extraction/Segmentation 27
       2.4.2 Shape Preprocessing 29
       2.4.3 Mode Decisions 29
   2.5 Summary 30
3. New Forms of Scalability in MPEG-4 33
   3.1 Object-Based Scalability 33
   3.2 Fine Granular Scalability 34
       3.2.1 FGS Coding with Adaptive Quantization (AQ) 38
   3.3 Hybrid Temporal-SNR Scalability with an all-FGS Structure 41
4. MPEG-4 Video Error Resilience 45
   4.1 Introduction 45
   4.2 MPEG-4 Video Transmission in Error-Prone Environment 46
       4.2.1 Overview 46
       4.2.2 Basic Principles in Error-Prone Video Transmission 48
   4.3 Error Resilience Tools in MPEG-4 53
       4.3.1 Introduction 53
       4.3.2 Resynchronization and Header Extension Code 53
       4.3.3 Data Partitioning 56

		4.3.4	Reversible Variable Length Codes............................57
		4.3.5	Intrarefresh 59
		4.3.6	New Prediction..61
	4.4	Streaming Protocols for MPEG-4 Video—A Brief Review 63
		4.4.1	Networks and Transport Protocols 63
		4.4.2	MPEG-4 Video over IP 63
		4.4.3	MPEG-4 Video over Wireless...............................66
5.	MPEG-4 Deployment: Ongoing Efforts....................................69

CHAPTER 1

Introduction

MPEG-4 (with a formal ISO/IEC designation ISO/IEC 14496) standardization was initiated in 1994 to address the requirements of the rapidly converging telecommunication, computer, and TV/film industries. MPEG-4 had a mandate to standardize algorithms for audiovisual coding in multimedia applications, digital television, interactive graphics, and interactive multimedia applications. The functionalities of MPEG-4 cover *content-based interactivity*, *universal access*, and *compression*, and a brief summary of these is provided in Table 1.1. MPEG-4 was finalized in October 1998 and became an international standard in the early months of 1999.

    The technologies developed during MPEG-4 standardization, leading to its current use especially in multimedia streaming systems and interactive applications, go significantly beyond the pure compression efficiency paradigm [1] under which MPEG-1 and MPEG-2 were developed. MPEG-4 was the first major attempt within the research community to examine object-based coding, i.e., decomposing a video scene into multiple arbitrarily shaped objects, and coding these objects separately and efficiently. This new approach enabled several additional functionalities such as region of interest coding, adapting, adding or deleting objects in the scene, etc., besides also having the potential to improve the coding efficiency. Furthermore, right from the outset, MPEG-4 was designed to enable universal access, covering a wide range of target bit-rates and receiver devices. Hence, an important aim of the standard was providing novel algorithms for scalability and error resilience. In this book, we use MPEG-4[1] as the backdrop to

---

[1]MPEG-4 has also additional components for combining audio and video with other rich media such as text, still images, animation, and 2-D and 3-D graphics, as well as a scripting language for elaborate

TABLE 1.1: Functionalities Within MPEG-4

Content-based interactivity	Content-based manipulation and bitstream editing without transcoding
	Hybrid natural and synthetic data coding
	Improved temporal random access within limited time frame and with fine resolution
Universal access	Robustness in error-prone environments including both wired and wireless networks, and high error conditions for low bit-rate video
	Fine-granular scalability in terms of content, quality, and complexity
	Target bit rates between 5 and 64 kb·s for mobile applications and up to 2 Mb/s for TV/film applications.
Compression	Improved coding efficiency
	Coding of multiple concurrent data streams, e.g., multiple views of video

describe the underlying principles and concepts behind some of these new technologies that continue to have significant impact in video coding and transmission applications.

We first present algorithms for content-based interactivity, focusing on coding and composition of objects with arbitrary shapes. We then describe technologies for universal access, such as Object-based Scalability and Fine Granularity Scalability (FGS). Finally, we discuss the use of MPEG-4 for multimedia streaming with a focus on error resilience.

programming. Recently, a new video coding standard within the MPEG-4 umbrella called MPEG-4 Part 10, which focuses primarily on compression efficiency, was also developed. Alternatively, in this book, we do not consider these, and focus on the MPEG-4 part 2 video standard.

We attempt to go beyond a simple description of what is included in the standard itself, and describe multiple algorithms that were evaluated during the course of the standard development. Furthermore, we also describe algorithms and techniques that lie outside the scope of the standard, but enable some of the functionalities supported by MPEG-4 applications. Given the growing deployment of MPEG-4 in multimedia streaming systems, we include a standard set of experimental results to highlight the advantages of these flexibilities especially for multimedia transmission across different kinds of networks and under varying streaming scenarios. Summarizing, this book is aimed at highlighting several key points that we believe have had a major impact on the adoption of MPEG-4 into existing products, and serve as an inspiration and basis for new, emerging standards and technologies. Additional information on MPEG-4, including a complete reference text, may be obtained from [2–5].

This book is organized as follows. Chapter 2 covers the coding of objects with arbitrary shape, including shape coding, texture coding, motion compensation techniques, and sprite coding. We also include a brief overview of some nonnormative parts of the standard such as segmentation, shape preprocessing, etc. Chapter 3 covers new forms of scalability in MPEG-4, including object-based scalability and FGS. We also include some discussion on hybrid forms of these scalabilities. In Chapter 4, we discuss the use of MPEG-4 for multimedia streaming and access. We describe briefly some standard error resilience and error concealment principles and highlight their use in the standard. We also describe packetization schemes used for MPEG-4 video. We present results of standard experiments that highlight the advantages of these various features for networks with different characteristics. Finally, in Chapter 5, we briefly describe the adoption of these technologies in applications and in the industry, and also ongoing efforts in the community to drive further deployment of MPEG-4 systems.

CHAPTER 2

Interactivity Support: Coding of Objects with Arbitrary Shapes

In this section we describe the support within MPEG-4 for coding objects with arbitrary shapes. In particular, there are three aspects that we focus on. We start by describing the decomposition of a particular video frame into multiple objects (with varying transparencies), object planes, etc. We then describe the algorithms for coding the shape, followed by algorithms to code the texture, including the use of motion compensation for such arbitrarily shaped objects. Toward the end of this chapter we describe Sprite Coding, an approach that encodes the background from multiple frames as one panoramic view (sprite). Finally, we describe some encoding considerations and additional algorithms that are not part of the MPEG-4 standard, but are required to enable object-based coding.

MPEG-4 supports the coding of multiple Video Object Planes (VOPs) as images of arbitrary shape[1] (corresponding to different objects) in order to achieve the desired content-based functionalities. A set of VOPs, possibly with arbitrary shapes and positions, can be collected into a Group of VOPs (GOV), and several GOVs can be collected into a Video Object Layer (VOL). A set of VOLs are collectively labeled a Video Object

---

[1] The coding of standard rectangular image sequences is supported as a special case of the VOP approach.

6   MPEG-4 BEYOND CONVENTIONAL VIDEO CODING

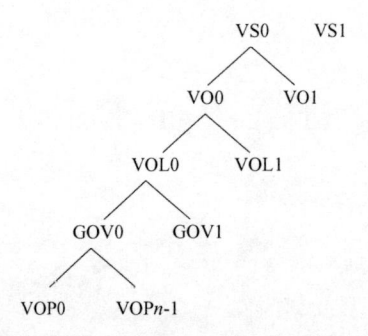

FIGURE 2.1: Object hierarchy within MPEG-4.

(VO), and sequence of VOs is termed a Visual object Sequence (VS). We show this hierarchy in Fig. 2.1.

An example of VOPs, VOLs, and VOs is shown in Fig. 2.2. In the figure, there are three VOPs, corresponding to the static background (VOP1), the tree (VOP2), and the man (VOP3). VOL1 is created by grouping VOP1 and VOP2 together, while VOL2 includes only VOP3. Finally, these different VOLs are composed into one VO.

Each VO in the scene is encoded and transmitted independently, and all the information required to identify each VO, and to help the compositor at the decoder insert these different VOs into the scene, is included in the bitstream.

It is assumed that the video sequence is segmented into a number of arbitrarily shaped VOPs containing particular content of interest, using online or offline segmentation techniques. As an illustration we show the segmented Akiyo sequence that consists

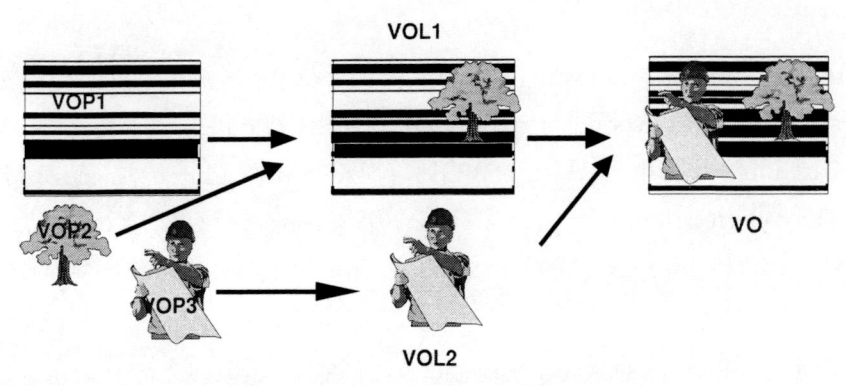

FIGURE 2.2: Video object planes, video object layers, and video objects.

FIGURE 2.3: Segmented Akiyo sequence with binary alpha map indicating shape and position of VOP1.

of a foreground object (VOP1) and a static background (VOP2) in Fig. 2.3. A binary alpha map is also coded to indicate the shape and the location of VOP1.

The alpha map indicates which pixels belong to the VOP1 (in this case the newscaster), and helps position it within the frame. MPEG-4 allows for overlapping and nonoverlapping VOPs. In general, MPEG-4 allows VOPs to have varying levels of transparency, and a grayscale alpha map (8 bit values with 0 representing completely transparent and 255 representing completely opaque) is used to represent the pixels of such VOPs. In Fig. 2.3 a binary alpha map is used to represent the pixels of VOP1 as it is completely opaque, i.e., completely occludes the background.

MPEG-4 builds upon previously defined coding standards like MPEG-1/2 and H.261/3 that use block-based coding schemes, and extends these to code VOPs with arbitrary shapes. To use these block-based schemes for VOPs with varying locations, sizes, and shapes, a shape-adaptive macroblock grid is employed. An example of an MPEG-4 macroblock grid for the foreground VOP in the Akiyo sequence, obtained from [6], is shown in Fig. 2.4.

A rectangular window with size multiple of 16 (macroblock size) in each direction is used to enclose the VOP and to specify the location of macroblocks within it. The window is typically located in such a way that the top-most and the left-most pixels of the VOP lie on the grid boundary. A shift parameter is coded to indicate the location of

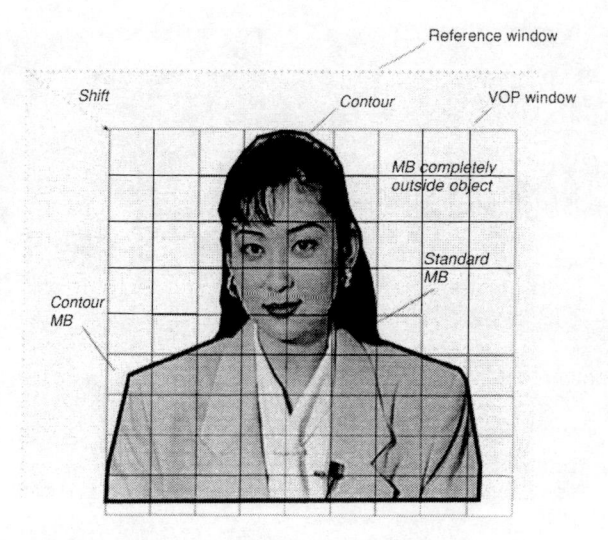

FIGURE 2.4: Shape adaptive macroblock grid for Akiyo foreground.

the VOP window with respect to the borders of a reference window (typically the image borders).

The coding of a VOP involves *adaptive shape coding* and *texture coding*, both of which may be performed with and without motion estimation and compensation. We describe shape coding in Section 2.1 and texture coding in Section 2.2.

2.1 SHAPE CODING

Two types of shape coding are supported within MPEG-4, binary alpha map coding and gray-scale alpha map coding. Binary shape coding is designed for opaque VOPs, while grayscale alpha map coding is designed to account for VOPs with varying transparencies.

2.1.1 Binary Shape Coding

There are three broad classes of binary shape coding techniques. Block-based coding and contour-based coding techniques code the shape explicitly, thereby encoding the alpha map that describes the shape of the VOP. In contrast, chroma keying encodes the shape of the VOP implicitly and does not require an alpha map. Different block-based and contour-based techniques were investigated within the MPEG-4 framework. These techniques are described in the following sections.

2.1.1.1 Block-based shape coding

Block-based coding techniques encode the shape of the VOP block by block. The shape-adaptive macroblock grid, shown in Fig. 2.4, is also superimposed on the alpha map, and each macroblock on this grid is labeled as a Binary Alpha Block (BAB). The shape is then encoded as a bitmap for each BAB. Within the bounding box, there are three different kinds of BABs:

a) those that lie completely inside the VOP;
b) those that lie completely outside the VOP; and
c) those that lie at boundaries, called boundary or contour BABs.

The shape does not need to be explicitly coded for BABs that lie either completely inside or completely outside the VOP, since these contain either all opaque (white) or all transparent (black) pixels, and it is enough to signal this, using the BAB type. The shape information needs to be explicitly encoded for boundary BABs, since these contain some opaque and some transparent pixels. Two different block-based shape coding techniques, context-based arithmetic encoding (CAE) and Modified Modified READ (MMR) coding, were investigated in MPEG-4, and these are described in Sections 2.1.1.1.1 and 2.1.1.1.2.

2.1.1.1.1 Context-Based Arithmetic Encoding. For boundary BABs, a context-based shape coder encodes the binary pixels in scan-line order (left to right and top to bottom) and exploits spatial redundancy with the shape information during encoding. A template of 10 causal pixels is used to define the context for predicting the shape value of the current pixel. This template is shown in Fig. 2.5.

Since the template extends two pixels above, to the right and to the left of the current pixel, some pixels of the BAB use context pixels from other BABs. When the current pixel lies in the top two rows or left two columns, corresponding context pixels

FIGURE 2.5: Context pixels for intracoding of shape.

from the BABs to the top and left are used. When the current pixel lies in the two right rows, context pixels outside the BAB are undefined, and are instead replaced by the value of their closest neighbor from within the current BAB. A context-based arithmetic coder is then used to encode the symbols. This arithmetic coder is trained on a previously selected training data set.

Intercoding of shape information may be used to further exploit temporal redundancies in VOP shapes. Two-dimensional (2-D) integer pixel shape motion vectors are estimated using a full search. The best matching shape region in the previous frame is determined by polygonal matching and is selected to minimize the prediction error for the current BAB. This is analogous to the estimation of texture motion vectors and is described in greater detail in Section 2.2.2. The shape motion vectors are encoded predictively (using their neighbors as predictors) in a process similar to the encoding of texture motion vectors. The motion vector coding overhead may be reduced by not estimating separate shape motion vectors, instead reusing texture motion vectors for shape information; however, this comes at the cost of worse prediction. Once the shape motion vectors are determined, they are used to align a new template to determine the contexts for the pixel being encoded. A context of nine pixels was defined for intercoding as shown in Fig. 2.6.

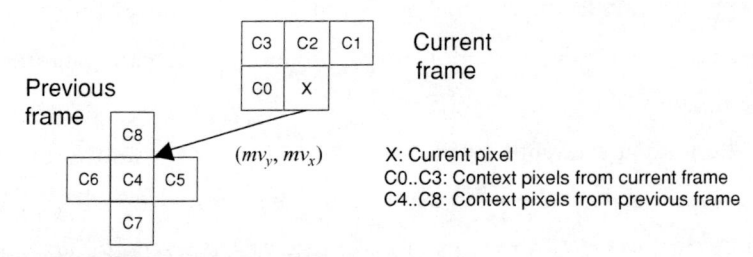

FIGURE 2.6: Context pixels for intercoding of shape.

In addition to four causal spatial neighbors, four pixels from the previous frame, at a location displaced by the corresponding shape motion vector (mv_y, mv_x), are also used as contexts. The decoder may further decide not to encode any prediction residue bits, and to reconstruct the VOP using only the shape information from previously decoded versions of the VOP, and the corresponding shape motion vectors.

To increase coding efficiency further, the BABs may be subsampled by a factor of 2 or 4; i.e., the BAB may be coded as a subsampled 8×8 block or as a 4×4 block. The subsampled blocks are then encoded using the techniques as above. This subsampling factor is also transmitted to the decoder so that it can upsample the decoded blocks appropriately. A higher subsampling factor leads to more efficient coding; however, this also leads to losses in the shape information and could lead to blockiness in the decoded shape. After experimental evaluations of subjective video quality, an adaptive nonlinear upsampling filter was selected by MPEG-4 for recovering the shape information at the decoder. The sampling grid for the pixels with both the subsampled pixel locations and the original pixel locations is shown in Fig. 2.7. Also shown is the set of pixels that are inputs (pixels at the subsampled locations) and outputs (reconstructed pixels at the original locations) of the upsampling filter.

Since the orientation of the shape in the VOP may be arbitrary, it may be beneficial to encode the shape top to bottom before left to right (for instance, when there are more vertical edges than horizontal edges). Hence the MPEG-4 encoder is allowed to transpose the BABs before encoding them. In summary, seven different modes may be used to code each BAB and these are shown in Table 2.1. More details on CAE of BABs may be obtained from [57].

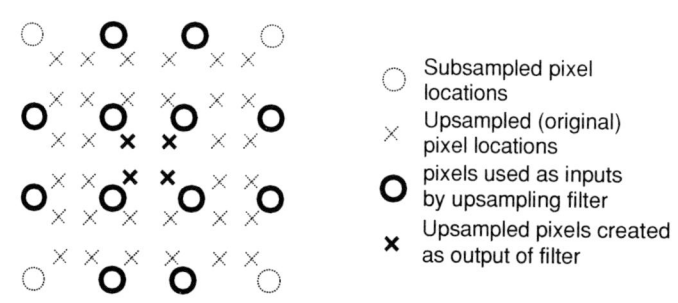

FIGURE 2.7: Location of samples for shape upsampling.

TABLE 2.1: Different Coding Modes Within the CAE Scheme

BAB TYPE	CODING MODE	DATA TO BE ENCODED	SUBSAMPLING	RASTER SCAN
Transparent Completely outside VOP	Intra/inter	Indicated using the BAB Type	Not used	Not used
Opaque Completely inside VOP	Intra/inter	Indicated using the BAB Type	Not used	Not used
Boundary Located at VOP boundary	Intra	The shape information is explicitly coded using intracontexts.	Factor of 1, 2, or 4	↗↘ or ↗↘
	Inter	Shape motion vectors are predictively coded, and shape information is coded using spatiotemporal contexts.	Factor of 1, 2, or 4	↗↘ or ↗↘

Inter without shape mvs	Texture motion vectors are reused, and shape information is coded using spatiotemporal contexts.	Factor of 1, 2, or 4 ↗ or ↗↘
Inter without prediction error	Shape motion vectors are predictively coded, and no prediction error is coded.	Not used	Not used
Inter without prediction error and shape mvs	Texture motion vectors are reused, and no prediction error is coded.	Not used	Not used

More details on CAE schemes may be obtained from [7-9].

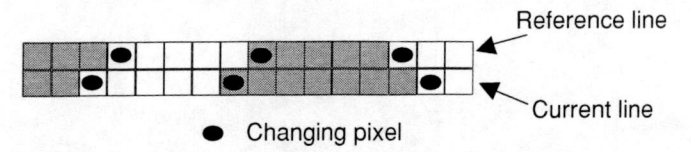

FIGURE 2.8: MMR coding used in the FAX standard.

2.1.1.1.2 Modified Modified READ (MMR) Shape Coding. In this shape coding technique [10], the BAB is directly encoded as a bitmap, using an MMR code (developed for the Fax standard). MMR coding encodes the binary data line by line. For each line of the data, it is necessary only to encode the positions of changing pixels (where the data change from black to white or vice versa). The positions of the changing pixels on the current line are then encoded relative to the positions of changing pixels on a reference line, chosen to be directly above the current line. An example of this is shown in Fig. 2.8.

After the current line is encoded, it may be used as a reference line for future lines. Like for the CAE scheme, BABs are coded differently on the basis of whether they are transparent, opaque, or boundary BABs. Only the type is used to indicate transparent and opaque BABs, while MMR codes are used for boundary BABs. In addition, motion compensation may be used to capture the temporal variation of shape, with full search used to determine the binary shape motion vectors, and the residual signal coded using the MMR codes. Each BAB may also be subsampled by a factor of 2 or 4, and this needs to be indicated to the decoder. Finally, the scan order may be vertical or horizontal based on the shape of the VOP.

2.1.1.2 Contour-Based Shape Coding

In contrast with block-based coding techniques, contour-based techniques encode the contour describing the shape of the VOP boundary. Two different contour-based techniques were investigated within the MPEG-4 framework, and these included vertex-based shape coding and baseline-based shape coding.

2.1.1.2.1 Vertex-Based Shape Coding. In vertex-based shape coding [11], the outline of the shape is represented using a polygonal approximation. A key component of vertex-based shape coding involves selecting appropriate vertices for the polygon. The placement

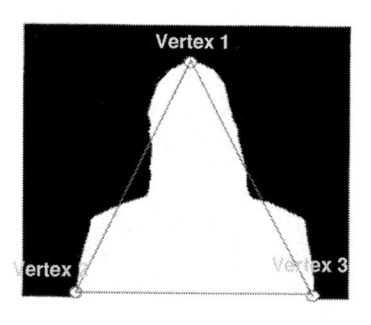

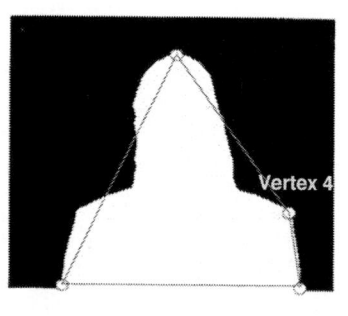

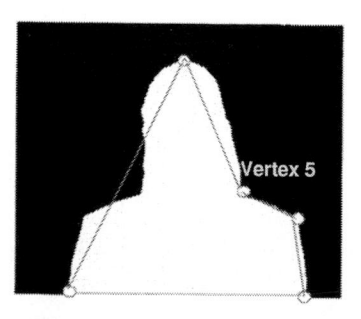

INTERACTIVITY SUPPORT: CODING OF OBJECTS WITH ARBITRARY SHAPES 15

FIGURE 2.9: Iterative shape approximation using polygons. Wherever the error exceeds the threshold, a new vertex is inserted.

of the vertices of the polygon controls the local variation in the shape approximation error. A common approach to vertex placement is as follows. The first two vertices are placed at the two ends of the main axis of the shape (the polygon in this case is a line). For each side of the polygon it is checked whether the shape approximation error lies within a predefined tolerance threshold. If the error exceeds the threshold, a new vertex is introduced at the point with the largest error, and the process is repeated for the newly generated sides of the polygon. This process is shown, for the shape map of the Akiyo foreground VOP, in Fig. 2.9.

Once the polygon is determined, only the positions of the vertices need to be transmitted to the decoder. In case lossless encoding of the shape is desired, each pixel on the shape boundary is labeled a vertex of the polygon. Chain coding [12, 13] is then used to encode the positions of these vertices efficiently. The shape is represented as a chain of vertices, using either a four-connected set of neighbors or an eight-connected set of neighbors. Each direction (spaced at 90° for the four-connected case or at 45° for the eight-connected case) is assigned a number, and the shape is described by a sequence of numbers corresponding to the traversing of these vertices in a clockwise manner. An example of this is shown in Fig. 2.10.

To further increase the coding efficiency, the chain may be differentially encoded, where the new local direction is computed relative to the previous local direction, i.e., by

16 MPEG-4 BEYOND CONVENTIONAL VIDEO CODING

Direct chain code:
−3,−3,4,4,−1,−2,−2,0,2
1,−1,−2,−2,2,1,1,2,2,4,3
Differential chain code:
−3,0,−1,0,3,−1,0,2,2,−1
−2,−1,0,4,−1,0,1,0,2,−1
Starting point

Four neighbors   Eight neighbors

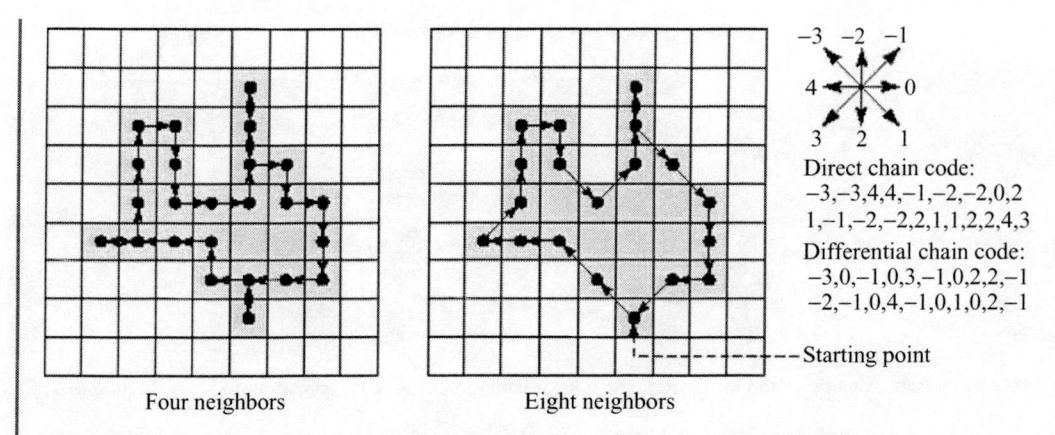

FIGURE 2.10: Chain coding with four- and eight-neighbor connectedness.

rotating the definition vectors so that 0 corresponds to the previous local direction. Finally, to capture the temporal shape variations, a motion vector can be assigned to each vertex.

2.1.1.2.2 Baseline-Based Shape Coding. Baseline-based shape coding [10] also encodes the contour describing the shape. The shape is placed onto a 2-D coordinate space with the X-axis corresponding to the main axis of the shape. The shape contour is then sampled clockwise and the y-coordinates of the shape boundary pixels are encoded differentially. Clearly, the x-coordinates of these contour pixels either decrease or increase continuously, and contour pixels where the direction changes are labeled turning points. The location of these turning points needs to be indicated to the decoder. An example of baseline-based coding for a contour is shown in Fig. 2.11.

In the figure, four different turning points are indicated, corresponding to when the X-coordinates of neighboring contour pixels change between continuously increasing, remaining the same, or continuously decreasing.

2.1.1.3 Chroma Key Shape Coding

Chroma key shape coding [14] was inspired from the blue-screen technique used by film and TV studios. Unlike the other schemes described, this is an implicit shaped coding technique. Pixels that lie outside the VOP are assigned a color, called a chroma key, not present in the VOP (typically a saturated color) and the resulting sequence of frames is encoded using a standard MPEG-4 coder. The chroma key is also indicated to the

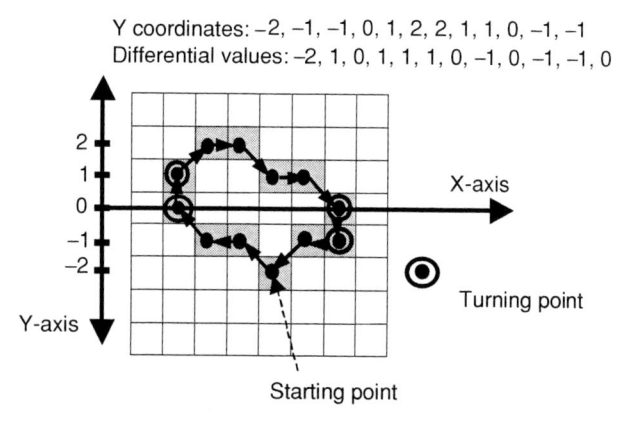

FIGURE 2.11: Baseline-based shape coding.

decoder where decoded pixels with color corresponding to the chroma-key are viewed as transparent. An important advantage of this scheme is the low computational and algorithmic complexity for the encoder and decoder. For simple objects like head and shoulders, chroma keying provides very good subjective quality. However, since the shape information is carried by the typically subsampled chroma components, this technique is not suitable for lossless shape coding.

2.1.1.4 Comparison of Different Shape Coding Techniques

During MPEG-4 Standardization, these different shape coding techniques were evaluated thoroughly in terms of their coding efficiency, subjective quality with lossy shape coding, hardware and software complexity, and their performance in scalable shape coders. Chroma keying was not included in the comparison as it is not as efficient as the other shape coding techniques, and the decoded shape topology was not stable, especially for complex objects. Furthermore, due to quantization and losses, the color of the key often bleeds into the object.

All the other shape coding schemes meet the requirements of the standard by providing, lossless, subjectively lossless and lossy shape coding. Furthermore, all these algorithms may be extended to allow scalable shape coding, bitstream editing, shape only decoding, and have support for low delay applications, as well as applications using error-prone channels.

The evaluation of the shape coders was performed in two stages. In the first stage, the contour-based schemes were compared against each other, and the block-based coding schemes were compared against each other, to determine the best contour-based shape coder, and the best block-based shape coder. In the second stage, the best contour-based coder was compared against the best block-based coder to determine the best shape coding scheme.

Among contour-based coding schemes, it was found that the vertex-based shape coder outperformed the baseline coder both in terms of coding efficiency for intercoding and in terms of computational complexity. Among the block-based coding schemes, the CAE coder outperformed the MMR coder for both intra- and intercoding of shape (both lossless and lossy). Hence, in the second stage, the vertex-based coder and the CAE were compared to determine the best shape coding technique. The results of this comparison, obtained from [7], are included in Table 2.2.

After the above-detailed comparison, the CAE was determined to have better performance[2] than the vertex-based coder and was selected to be part of the standard.

2.1.2 Grayscale Shape Coding

Grayscale alpha map coding is used to code the shape and transparency of VOPs in the scene. Unlike in binary shape coding, where all the blocks completely inside the VOP are opaque, in grayscale alpha map coding, different blocks of the VOP may have different transparencies. There are two different cases of grayscale alpha map coding.

2.1.2.1 VOPs with Constant Transparency

In this case, grayscale alpha map coding degenerates to binary shape coding; however, in addition to the binary shape, the 8 bit alpha value corresponding to the transparency of the VOP also needs to be transmitted. In some cases, the alpha map near the VOP boundary is filtered to blend the VOP into the scene. Different filters may be applied to a strip of width up to three pixels inside the VOP boundary, to allow this blending. In such cases, the filter coefficients also need to be transmitted to the decoder.

---

[2]Recent experiments have shown that chain coding performed on a block-by-block basis performs comparably with CAE for intracoding.

TABLE 2.2: Comparison Between CAE and Vertex-Based Shape Coding

	CAE	**VERTEX BASED**
Coding efficiency: Intra Lossless		7.8% lower data rate
Coding efficiency: Inter Lossless	20.5% lower data rate	
Coding efficiency: Inter Lossy	Better at small distortions	Better at large distortions
Scalability overhead for three layers (layer three lossless)	30–50% of lossless one layer rate for predictive coding	No optimized results for inter coding
Delay	Slightly lower	
Hardware implementation complexity	Decoding on chip without access to external memory	Huffman decoder smaller than arithmetic decoder, however, required random access to external memory
Software implementation complexity	No optimized coder was available; however, the nonoptimized code had similar performance for both algorithms	

2.1.2.2 VOPs with Varying Transparency

For VOPs with arbitrary transparencies, the shape coding is performed in two steps. First the outline of the shape is encoded using binary shape coding techniques. In the second step, the alpha map values are viewed as luminance values and are coded using padding, motion compensation, and DCT. More details on padding are included in Section 2.2.1.1.

2.2 TEXTURE CODING

2.2.1 Intracoding

The texture is coded for each macroblock within the shape adaptive grid, using the standard 8 × 8 DCT. No texture information is encoded for 8 × 8 blocks that lie completely outside the VOP. The regular 8 × 8 DCT is used to encode the texture of blocks that lie completely inside the VOP. The texture of boundary blocks, which have some transparent pixels (pixels that lie outside the VOP boundary), is encoded using two different techniques: padding followed by 8 × 8 DCT and shape-adaptive DCT.

2.2.1.1 Padding for Intracoding of Boundary Blocks

When applying the 8 × 8 DCT to the boundary blocks, the transparent pixels need to be assigned YUV values. In theory, these transparent pixels can be given arbitrary values, since they are discarded at the decoder anyway. Values assigned to these transparent pixels in no way affect conformance to the standard. However, assigning arbitrary values to these pixels can lead to large and high-frequency DCT coefficients, and lead to coding inefficiencies. It was determined during the MPEG-4 core experiments that simple low-pass extrapolation is an efficient way to assign values to these pixels. This involves, first, replicating the average of all the opaque pixels in the block, across the transparent pixels. Finally, a filter is applied recursively to each of the transparent pixels in the raster scan order, where each pixel is replaced by the average of its four neighbors.

$$y(p,q) = \frac{1}{4}[y(p-1,q) + y(p,q-1) + y(p,q+1) + y(p+1,q)]$$

with (p, q) the location of the current pixel. This is shown in Fig. 2.12.

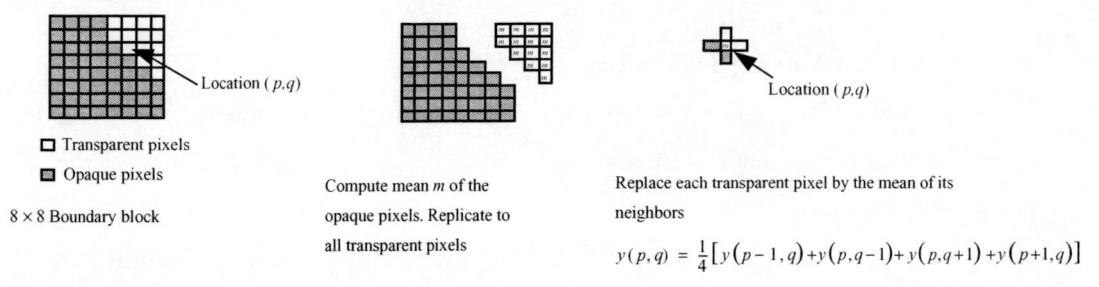

FIGURE 2.12: Padding for texture coding of boundary blocks.

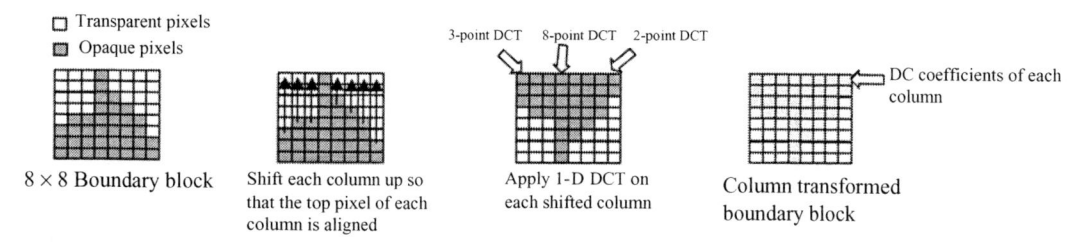

FIGURE 2.13: Shape-adaptive column DCT for boundary block.

2.2.1.2 Shape Adaptive DCT

The shape adaptive DCT is another way of coding the texture of boundary blocks and was developed in [58] based on earlier work in [59]. The standard 8 × 8 DCT is a separable 2-D transform that is implemented as a succession of two one-dimensional (1-D) transforms first applied column by column, and then applied row by row.[3] However, in a boundary block, the number of opaque pixels varies in each column and row. Hence, instead of performing a succession of 8-point 1-D DCTs, we may perform a succession of variable length n-point DCTs ($n = 8$) corresponding to the number of opaque pixels in the row/column. Before we apply this variable length DCT to each row/column, the pixels need to be aligned so that transform coefficients corresponding to similar frequencies are present in similar positions. We first illustrate the use of variable length DCTs on the columns of a sample boundary block in Fig. 2.13.

Once the columns are transformed, these transform coefficients are transformed row by row to remove any redundancies in the horizontal direction. Again, the rows are first aligned, and the process is shown in Fig. 2.14.

Finally, the coefficients are quantized and encoded in a manner identical to the coefficients obtained after the 8 × 8 2-D DCT. At the decoder first the shape is decoded, and then the texture can be decoded by shifting the received coefficients appropriately and inverting the variable length DCTs. Although this scheme is more complex than the padding for texture coding, it shows 1–3 dB gains in the decoded video quality.

---

[3]The 1-D transforms may also be applied first on the rows and then on the columns.

22 MPEG-4 BEYOND CONVENTIONAL VIDEO CODING

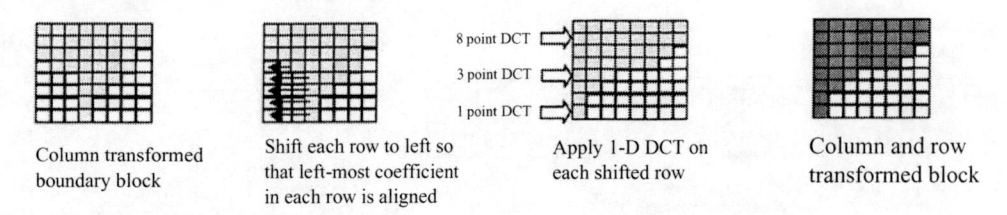

FIGURE 2.14: Shape-adaptive row DCT performed after column DCT.

2.2.2 Intercoding

Motion estimation and compensation techniques are used for each macroblock, within the shape adaptive grid, to remove temporal redundancies. In general, the motion estimation and compensation techniques used in MPEG-4 may be viewed as an extension of those used in H.263 [60] and MPEG-2. Different estimation and compensation techniques are used for different types of macroblocks, and these are shown in Fig. 2.15.

Clearly, no matching is performed for macroblocks that lie completely outside the VOP. For macroblocks completely inside the VOP, conventional block matching, as in

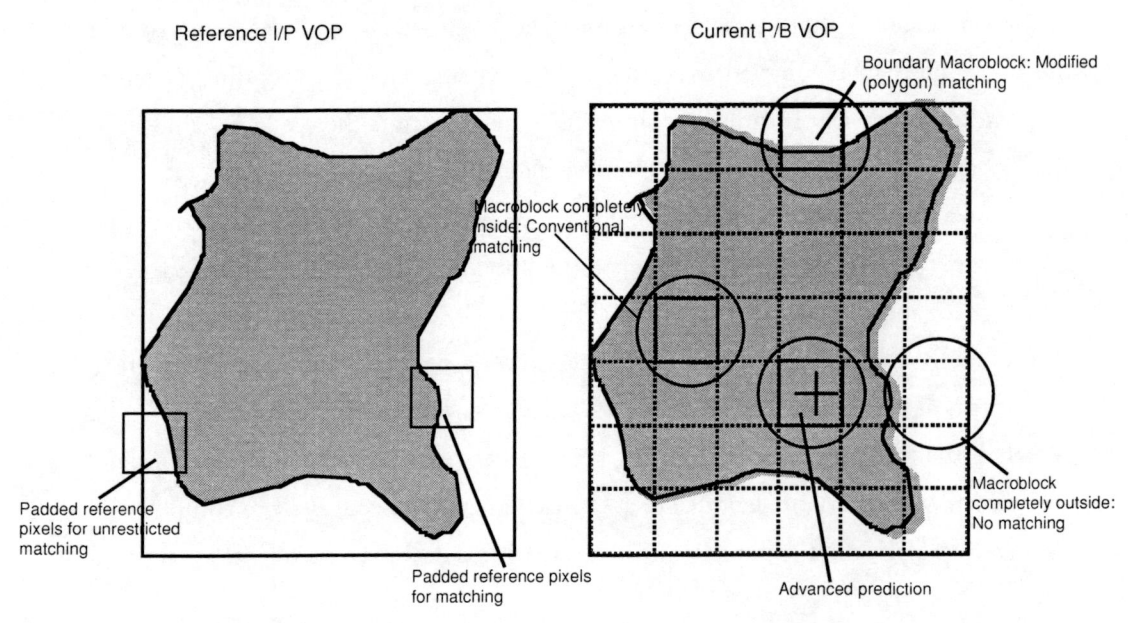

FIGURE 2.15: Motion estimation and compensation techniques for different macroblocks.

FIGURE 2.16: Padded reference VOP used for motion compensation.

previous video coding standards, is performed. The prediction error is determined and coded along with the motion vector (called the texture motion vector). An advanced motion compensation mode is also supported within the standard. This advanced mode allows for the use of overlapped block motion compensation (OBMC) as in the H.263 standard, and also allows for estimation of motion vectors for 8×8 blocks.

To estimate motion vectors for boundary macroblocks, the reference VOP is extrapolated using the image padding technique described in Section 2.2.1.1. This padding may extrapolate the VOP pixels both within and outside the bounding rectangular window, since the search range can include regions outside the bounding window, for unrestricted motion vector search. An example of the padded reference VOP from the Akiyo sequence is shown in Fig. 2.16.

Once the reference VOP is padded, a polygonal shape matching technique is used to determine the best match for the boundary macroblock. A polygon is used to define the part of the boundary macroblock that lies inside the VOP, and when block matching is performed, only pixels within this polygon are considered. For instance, when computing the sum of absolute difference (SAD) during block matching, only differences for pixels that lie inside the polygon are considered.

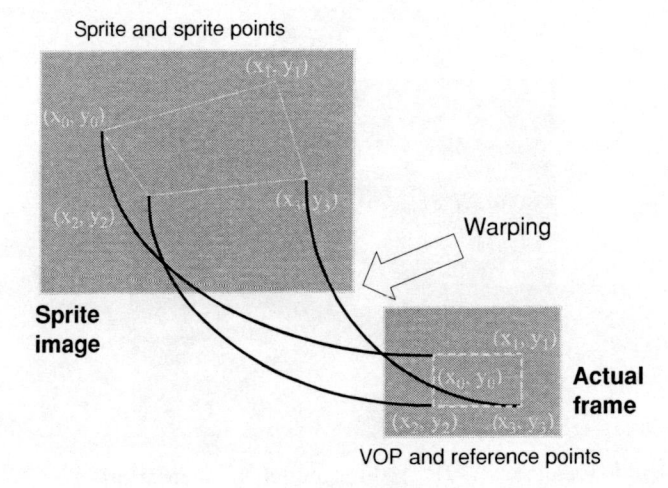

FIGURE 2.17: Warping of the sprite to reconstruct the background object.

MPEG-4 supports the coding of both forward-predicted (P) and bidirectionally predicted (B) VOPs. In the case of bidirectional prediction, the average between the forward and backward best matched regions is used as a predictor. The texture motion vectors are predictively coded using standard H.263 VLC code tables.

2.3 SPRITE CODING

A sprite, also referred to as a mosaic, is an image describing a panoramic view of a video object visible throughout a video segment. As an example, a sprite for the background object generated from a scene with a panning camera will contain all the visible pixels[4] during that scene. To generate sprite images, the video sequence is partitioned into a set of subsequences with similar content (using scene cut detection techniques). A different background sprite image is generated for each subsequence. The background object is segmented in each frame of the subsequence and warped to a fixed coordinate system after estimating its motion. For MPEG-4 content, the warping is typically performed by assigning 2-D motion vectors to a set of points on the object labeled *reference points*. These reference points are shown in Fig. 2.17 as the vertices of the polygon.

---

[4]Not all pixels of the background object may be visible due to the presence of a foreground object with its own motion.

FIGURE 2.18: Background sprite for the Stefan sequence.

This process of warping corresponds to the application of an affine transformation to the background object, corresponding to the estimated motion. Once the warped background images are obtained from the frames in the subsequence, the information from them is combined into the background sprite image, using median filtering or averaging operations. An example background sprite, for the Stefan sequence, is shown in Fig. 2.18.

Sprite images typically provide a concise representation of the background in a scene. Since the sprite contains all parts of the background that were visible at least once, the sprite can be used for the reconstruction or for the predictive coding of the background object. Hence, if the background sprite image is available at the decoder, the background of each frame in the subsequence can be generated from this, using the inverse of the warping procedure used to create the sprite. MPEG-4 allows the reconstruction of the background objects from the sprite images using a set of 2–8 global motion parameters. There are two different sprite coding techniques supported within MPEG-4, static sprite coding and dynamic sprite coding. In static sprite coding, the sprite is generated off-line prior to the encoding and transmitted to the decoder as first frame of the sequence. The background sprite image itself is treated as a VOP with arbitrary shape and coded using techniques described in Sections 2.1 and 2.2. At the decoder, the decoded sprite image is stored in a sprite buffer. In each consecutive frame, only the camera parameters, required for the generation of the background from the sprite, are transmitted. The moving foreground object is transmitted separately as an

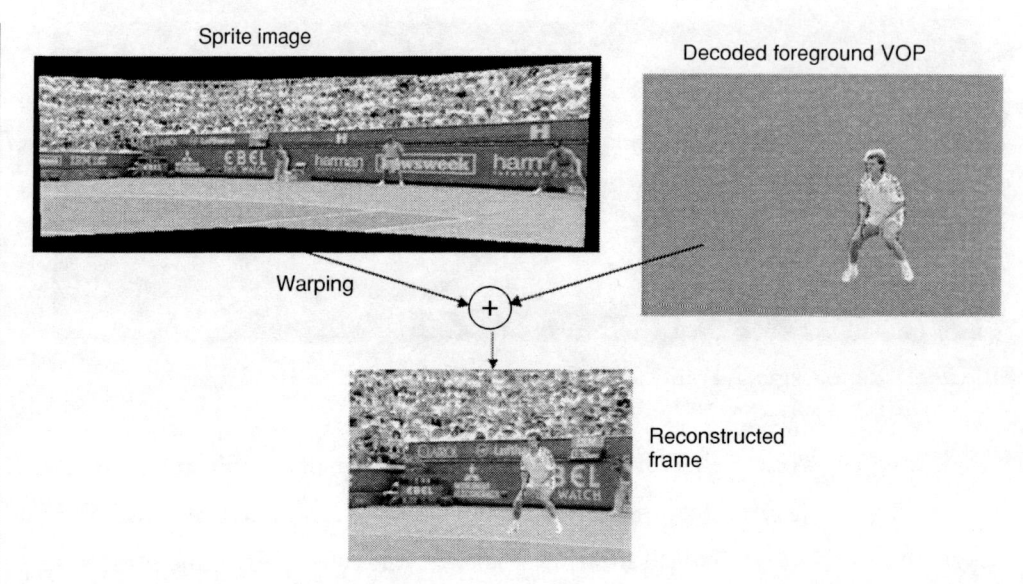

FIGURE 2.19: Combining decoded background object (from warped sprite) and foreground to obtain decoded frame.

arbitrary-shape video object. Finally, the decoded foreground and background objects may be combined to obtain the reconstructed sequence, and an example is shown in 2.19.

Since the static sprites may be very large images, transmitting the entire sprite as the first frame might not be suitable for low delay applications. Hence, MPEG-4 also supports a low-latency mode, where it is possible to transmit the sprite in multiple smaller pieces over consecutive frames or to build up the sprite at the decoder progressively.

Dynamic sprites, on the other hand, are not computed offline, but are generated on the fly from the background objects, using global motion compensation (GMC) techniques. Short-term GMC parameters are estimated from successive frames in the sequence and used to warp the sprite at each step. The image generated from this warped sprite is used as a reference for the predictive coding of the background in the current frame, and the residual error is encoded and transmitted to the decoder.[5] The sprite is updated by blending from the reconstructed image. This scheme avoids the overhead of transmitting a large sprite image at the start of the sequence; however, as the updating

---

[5]The residual image needs to be always encoded as otherwise there will be prediction drift between the encoder and the decoder.

of the sprite at every time step also needs to be performed at the decoder, it can increase the decoder complexity. More details on sprite coding may be obtained from [15–17].

2.4 ENCODING CONSIDERATIONS

This section covers information that is not part of the standard, but is useful to know if implementing systems based on the standard. This includes segmenting the arbitrarily shaped objects from a video scene, preprocessing shape, and coding mode decisions. The coding performance of MPEG-4 depends heavily on the algorithms used for these steps; however, these are difficult problems, and the design of optimal algorithms for them is still an area of open research.

2.4.1 Shape Extraction/Segmentation

A key goal of segmentation is the detection of spatial or temporal transitions and discontinuities in the video signal that partition it into the underlying multiple objects. The detection of these multiple objects is simplified if we have access to the content creation process; however, in general the task of segmentation is posed as problem of estimating object boundaries after the content has already been created. This makes it a difficult problem to solve, and the state-of-the-art needs to be considerably improved to robustly deal with generic images and video sequences.

The typical segmentation process consists of three major steps, *simplification, feature extraction*, and *decision*, and these are shown in greater detail in Fig. 2.20.

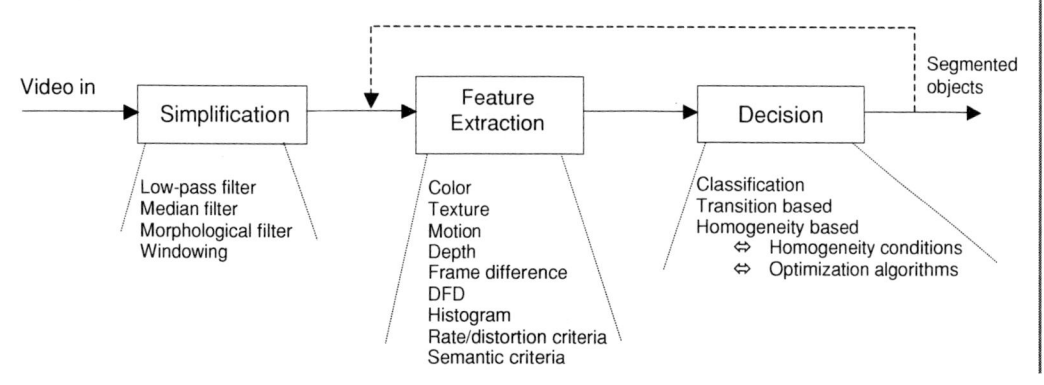

FIGURE 2.20: Major steps in the segmentation process.

Simplification is a preprocessing stage that helps remove irrelevant information from the content and results in data that are easier to segment. For instance, complex details may be removed from textured areas without affecting the object boundaries. Different techniques such as low-pass filtering, median filtering, windowing, etc., are used during this process.

Once the content is simplified, features are extracted from the video. Appropriate features are selected on the basis of the type of homogeneity expected within the partitions. These features describe different aspects of the underlying content and can include information about the texture, the motion, the depth or the displaced frame difference (DFD), or even semantic information about the scene. Oftentimes an iterative procedure is used for segmentation where features are reextracted from the previous segmented result to improve the final result. In such cases a loop is introduced in the segmentation process, and this is shown in Fig. 2.20 as a dotted line.

Finally, the decision stage consists of analyzing the feature space to partition the data into separate areas with distinct characteristics in terms of the extracted features. Some common techniques used within the decision process include classification techniques, transition-based techniques, and homogeneity-based techniques. An example of segmentation using homogeneity-based techniques in conjunction with the use of texture and contour features, obtained from [18], is shown in Fig. 2.21.

More details on segmentation and shape extraction may be obtained from [18].

FIGURE 2.21: Segmentation example using homogeneity and texture and contour features: (a) original image, (b) segmented with low weight for contour features, and (c) high weight for contour features.

2.4.2 Shape Preprocessing

Due to imperfections in the capture process, the binary alpha map of the video (the component that defines the transparencies of the pixels) may be noisy. The impact of a noisy alpha map can be significant in terms of both the shape and the texture bit-rate. First, a noisy alpha map can reduce the number of coherent alpha blocks, i.e., blocks with a constant transparency. Hence, instead of coding one value for the block, the transparency of each pixel in the block needs to be coded, leading to an increased shape bit rate. Second, the presence of noise can reduce the number of transparent pixels, requiring the coding of a larger amount of texture information, and thereby increasing the texture bit rate. Shape information needs to be preprocessed in order to remove any noise. Connected morphological and median filters have been proposed for the preprocessing of shape information for noise removal.

In addition to preprocessing for noise removal, the location of the shape adaptive grid may be adjusted to minimize the number of macroblocks to be coded, and also the number of nontransparent blocks, thereby reducing the bits for both the texture information and the motion vectors.

2.4.3 Mode Decisions

There are many different mode decisions that need to be made at different granularities, during the coding of shape and texture for arbitrarily shaped VOPs. At the VOP level, it must first be decided whether to code the VOP in an intramode or an intermode. Although intercoding is typically more efficient than intracoding, when the VOPs are not rigid and the motion is large and random, intracoding can be more efficient than intercoding. Intracoding can also reduce the propagation of errors when the video is transmitted over lossy networks. Similarly, the coding mode for each macroblock/BAB needs to be determined from among the seven modes listed in Table 2.1. Other mode decisions include deciding whether the BAB should be subsampled or not, and whether it should be coded using a vertical, or a horizontal raster scan. These mode decisions are not a normative part of the standard and constitute encoder optimizations corresponding to different application requirements. These mode decisions need to be implemented

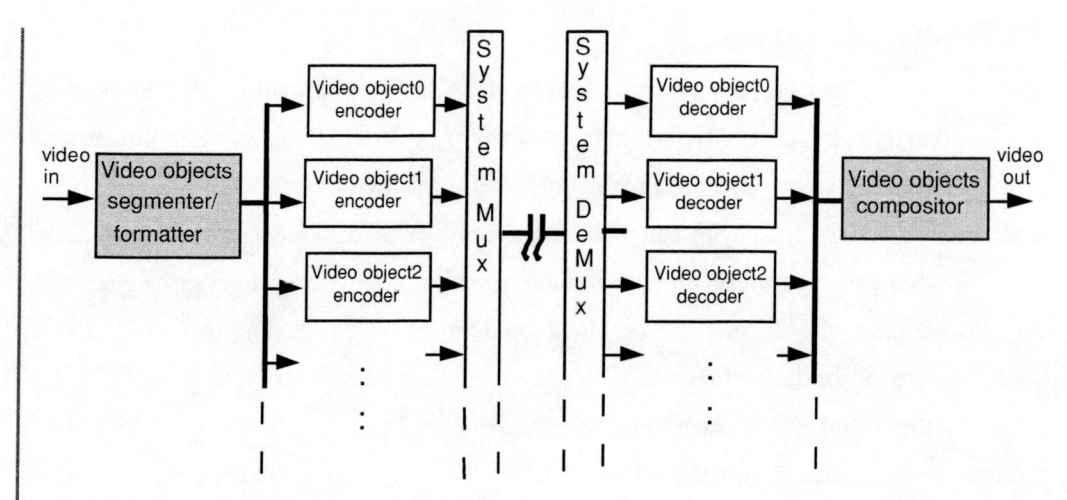

FIGURE 2.22: MPEG-4 encoder decoder pair for coding VOPs with arbitrary shape.

keeping in mind the bit rate, the distortion, the complexity, the user requirements (e.g., allowable shape distortion threshold), and the error resilience, or a combination of any of these factors. The design of appropriate mode decisions forms an interesting area of research, and there is a large amount of literature available on the topic.

2.5 SUMMARY

The overall structure of an MPEG-4 encoder-decoder pair is as shown in Fig. 2.22. The segmenter and the compositor are shown as pre- and postprocessing modules, and are not part of the encoder decoder pair. As may be observed, each VO is encoded and decoded separately. The bitstreams for all these VOs are multiplexed into a common bitstream. Also included is the information to composit (compose) these objects into the scene. The decoder can then decode appropriate parts of the bitstream and composit the different objects into the scene.

The block diagram of the VOP decoder is shown in greater detail in Fig. 2.23. The decoder first demultiplexes information about the shape, the texture, and the motion from the received bitstream. There are different basic subcoders for shape and for texture, both of which may be intracoded or intercoded. The techniques described in Sections 2.1

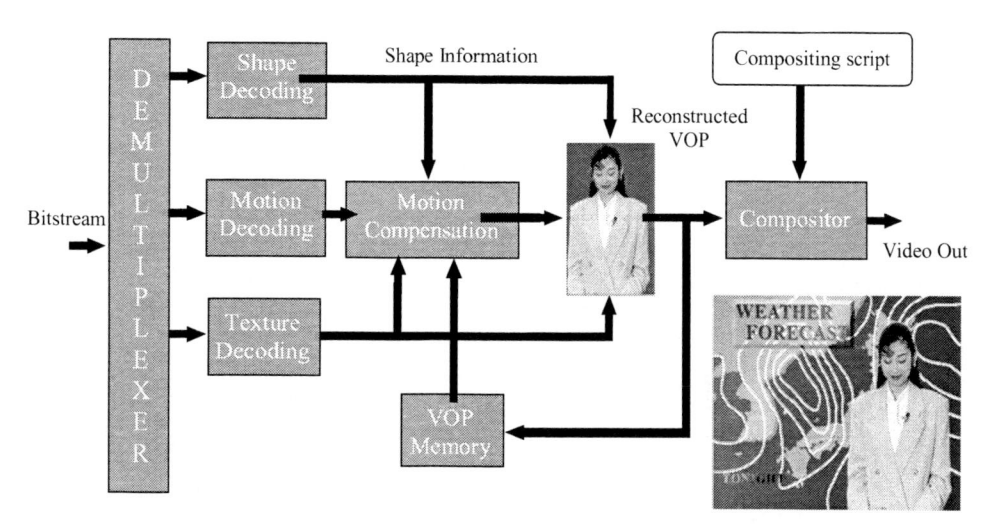

FIGURE 2.23: VOP decoder for objects with arbitrary shapes.

and 2.2 are used to decode the shape and the texture appropriately, and then these are combined to reconstruct the VOP. Once the VOP is decoded, it may then be composited into the scene.

CHAPTER 3

New Forms of Scalability in MPEG-4

Scalability is essential to address the requirements of various emerging video streaming applications, such as video transmission over low bit-rate wireless networks or the Internet. Previously proposed scalability modes included spatial and temporal scalability that are part of the MPEG-2 standard. MPEG-4 introduced two new forms of scalability: Object-based scalability and fine granular scalability (FGS). In this section we describe these in more detail. Furthermore, MPEG-4 also enabled hybrid scalability, i.e., combining different scalability modes, and we also describe Hybrid temporal-SNR scalability.

3.1 OBJECT-BASED SCALABILITY

MPEG-4 allows for content-based functionalities, i.e., the ability to identify and selectively decode and reconstruct video content of interest. This feature provides a simple mechanism for interactivity and content manipulation in the compressed domain, without the need for complex segmentation or transcoding at the decoder. This MPEG-4 tool allows emphasizing relevant objects within the video by enhancing their quality, spatial resolution, or temporal resolution. Using the object-based scalability, optimum trade-off between spatial/temporal/quality resolution based on scene content can be achieved. In Fig. 3.1 we show an example where one particular object (shown as an ellipse) is selectively enhanced to improve its quality.

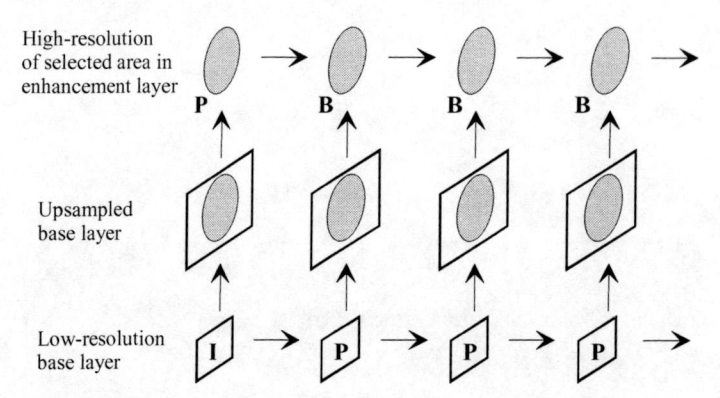

FIGURE 3.1: Illustration of object-based scalable coding in MPEG-4.

The object-based scalability can be employed using both arbitrary-shaped objects and rectangular blocks in a block-based coder.

3.2 FINE GRANULAR SCALABILITY

FGS [19–23] consists of a rich set of video coding tools that support quality (i.e., SNR), temporal, and hybrid temporal-SNR scalabilities. Furthermore, FGS is simple and flexible in supporting unicast and multicast video streaming applications over IP [23].

As shown in Fig. 3.2, the FGS framework requires two encoders, one for the *base layer* and the other for the *enhancement layer*. The base layer can be compressed using DCT-based MPEG-4 tools, as described in the preceding sections.

In principle, the FGS enhancement-layer encoder can be based on any fine-granular coding method. However, due to the fact that the FGS base layer is coded using DCT coding, employing embedded DCT method (i.e. coding data bitplane by bitplane) for compressing the enhancement layer is a sensible option [22].

The enhancement layer consists of residual DCT coefficients that are obtained as shown in Fig. 3.2 by subtracting the dequantized DCT coefficients of the base layer from the DCT coefficients of the original or motion compensated frames.

Once the enhancement layer FGS residual coefficients are obtained, they are coded bitplane by bitplane. During this process, different optional Adaptive Quantization

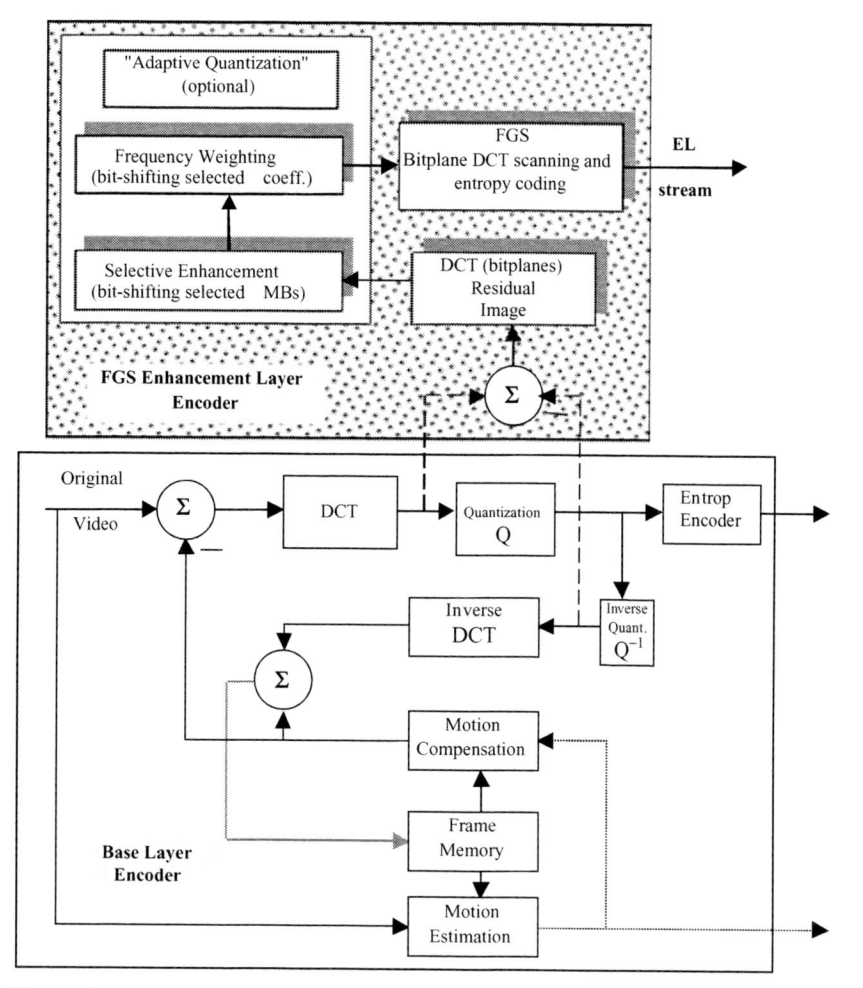

FIGURE 3.2: Obtaining and coding the enhancement layer.

mechanisms such as Selective Enhancement and Frequency Weighting can be used [19]. These are described in more detail in Section 3.2.1.

Each DCT FGS-residual frame consists of $N_{BP} = \log_2(|C|_{max}) + 1$ bitplanes, where $|C|_{max}$ is the maximum DCT (magnitude) value of the residual coefficients under consideration.[1] After identifying $|C|_{max}$ and the corresponding N_{BP}, the FGS enhancement-layer encoder scans the residual frame macroblock by macroblock (MBs),

---

[1] In the FGS MPEG-4 standard, three parameters are used for the number-of-bitplanes variable N_{BP}: $N_{BP}(Y)$, $N_{BP}(U)$, and $N_{BP}(V)$ for the luminance and chroma components of the video signal.

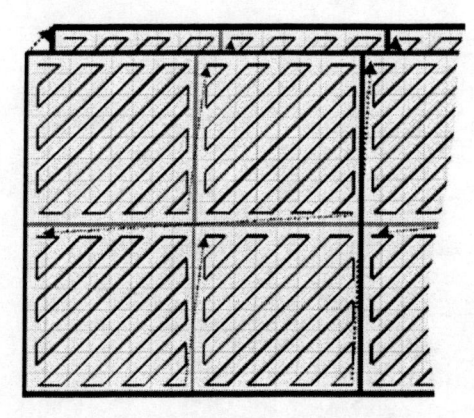

FIGURE 3.3: Scanning of residual coefficients for each bitplane.

where each MB includes four 8×8 luminance (Y) blocks and two chroma blocks (U and V), starting from the most significant bitplane BP(1) and ending at the least significant bitplane[2] BP(N_{BP}). Within each block the coefficients are scanned in the traditional zig-zag scanning method as shown in Fig. 3.3 and the corresponding values (zeros and ones) are then entropy coded. At bitplane B, for any coefficient, this value corresponds to the bit in location B (going from LSB to MSB) in the binary representation of the coefficient magnitude. Hence, while scanning at bitplane 3, a coefficient with magnitude 12 (1100) will have corresponding value 1 while a coefficient with magnitude 11 (1011) will have corresponding value 0.

Run-length codes are used for (lossless) entropy-coding of the zeros and ones in each 8×8 bitplane block [21, 22]. This process generates variable length codes that constitute the FGS-compressed bitstream. A special "all-zero blocks" code is used when all six bitplane blocks (within a given bitplane macroblock) do not have any bits with a value of 1.

At the receiver side, the FGS bitstream is first decoded by a Variable Length Decoder (VLD) as shown in Fig. 3.4.

Due to the embedded nature of the FGS stream, the VLD regenerates the DCT residual bitplanes starting from the most significant bitplane toward the least significant

---

[2]Alternatively, the encoder may stop encoding the residual signal if the desired maximum bitrate is reached.

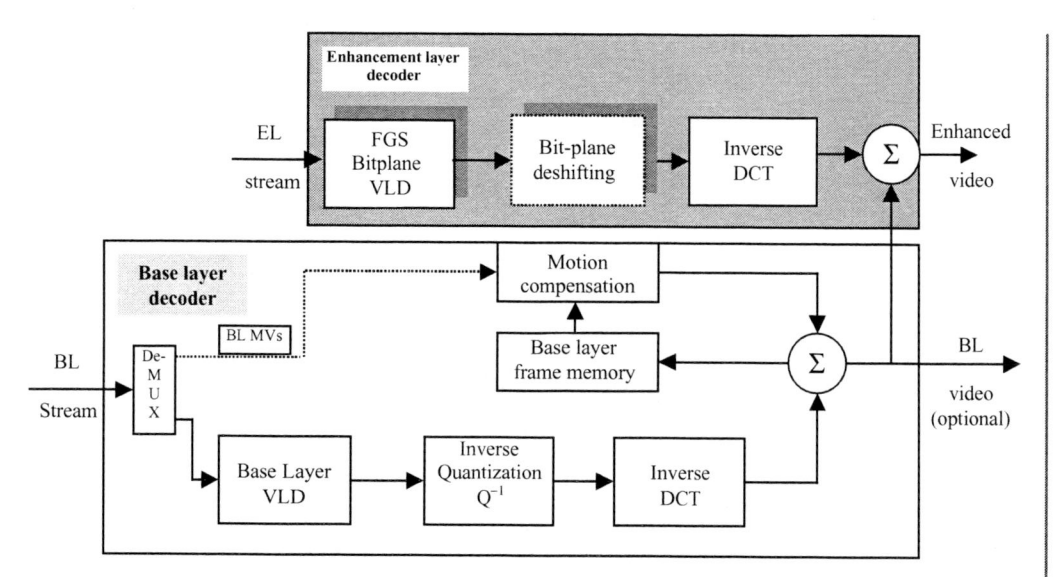

FIGURE 3.4: FGS SNR decoder. The FGS decoder includes bitplane deshifting to compensate for the two "Adaptive Quantization" encoding tools: Selective Enhancement and Frequency Weighting.

one. Moreover, due to the type of scanning used by the FGS encoder (Fig. 3.3), it is possible that the decoder does not receive all of the bitplane blocks that belong to a particular bitplane. Any bitplane block not received by the decoder can be filled with zero values.[3] The resulting DCT residual is then inverse-transformed to generate the SNR residual pixels. These residual pixels are then added to the base-layer decoder output to generate the final enhanced scalable video.

In summary, the basic SNR FGS codec employs embedded DCT variable length encoding and decoding operations that resemble the ones used in typical DCT-based standards. In previous standards (including MPEG-4 base layer), the DCT coefficients are coded with (run-length, amplitude) type of codes, whereas with FGS the bitplane ones-and-zeros are coded with (run-length) codes since the "amplitude" is

---

[3]For an "optimal" reconstruction (in a mean-square-error sense) of the DCT coefficients, one forth (1/4) of the received quantization step size is added. For example, if the decoder receives only the MSB of a coefficient (with a value x, where $x=0$ or 1), this coefficient is reconstructed using the value $x01000$ (i.e., instead of $x0000...$).

always one. More information about the VLC codes used by FGS may be obtained from [21].

3.2.1 FGS Coding with Adaptive Quantization (AQ)

AQ is a very useful coding tool for improving the visual quality of transform-coded video. It is normally achieved through a quantization matrix that defines different quantization step sizes for the different transform coefficients within a block (prior to performing entropy coding on these coefficients). For example, the DC coefficient and other "low frequency" coefficients normally contribute more to the visual quality, and consequently small step sizes are used for quantizing them. AQ can also be controlled from one macroblock to another through a quantization factor whose value varies on a macroblock-by-macroblock basis. These AQ tools have been employed successfully in the MPEG-2 and MPEG-4 (base-layer) standards.

Performing AQ on bitplane signals consisting of only ones and zeros has to be achieved through a different (yet conceptually similar) set of techniques. FGS-based AQ is achieved through *bitplane shifting* of (a) selected macroblocks within an FGS enhancement layer frame and/or (b) selected coefficients within the 8 × 8 blocks. Bitplane shifting is equivalent to multiplying a particular set of coefficients by a power-of-2 integer. For example, let assume that the FGS encoder wishes to emphasize a particular macroblock k within an FGS frame. All blocks within this selected macroblock k can be multiplied[4] by a factor $2^{n_{se}(k)}$. Therefore, the new value $c'(i; j; k)$ of a coefficient i of block j (within macroblock k) is:

$$c'(i, j, k) = 2^{n_{se}(k)} \cdot c(i, j, k)$$

where $c(i, j, k)$ is the original value of the coefficient. This is equivalent to *up-shifting* the set of coefficients $[c(i; j; k), i = 1, 2, \ldots 64]$ by $n_{se}(k)$ bitplanes relative to other coefficients that belong to other macroblocks. An example of this is illustrated in Fig. 3.5. This type of adaptive-quantization tool is referred to as *Selective Enhancement* [24] since

---

[4]Throughout the remainder of this section we will use the words *shifted*, *multiplied*, and *upshifted* interchangeably.

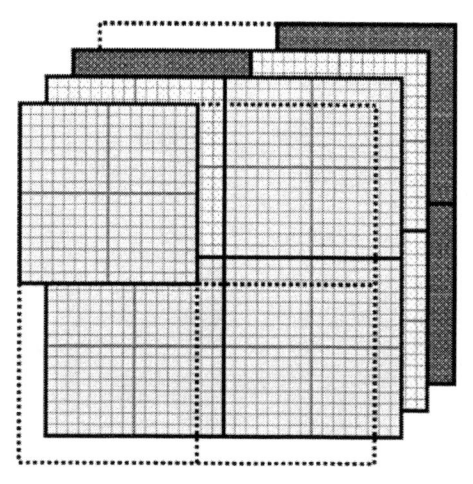

FIGURE 3.5: Example illustrating the use of the selective enhancement AQ tool.

through this approach selected macroblocks within a given frame can be enhanced relative to other macroblocks within the same frame. Of course, such upshifting generates a new bitplane when compared to the original number of bitplanes.

In addition to performing bitplane shifting on selected macroblocks, FGS allows bitplane shifting of selected DCT coefficients, as shown in Fig. 3.6. Therefore, one can define a *frequency weighting* matrix where each element of the matrix indicates the number of bitplanes $n_{fw}(i)$ that the ith coefficient should be shifted by. Again, this is

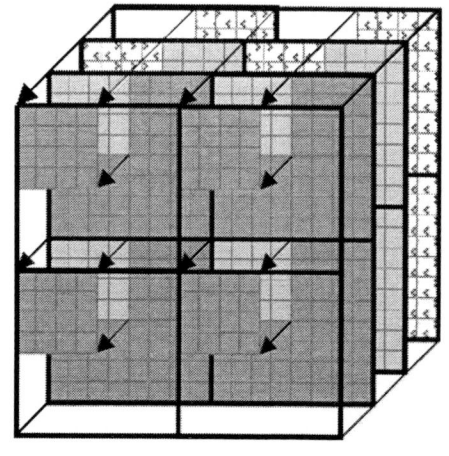

FIGURE 3.6: Example FGS frequency weighting AQ tool.

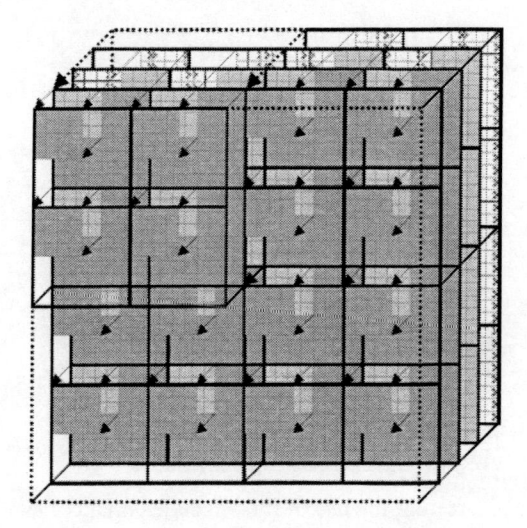

FIGURE 3.7: Example illustrating use of both selective enhancement and frequency weighting.

equivalent to multiplying the DCT coefficients by an integer:

$$c'(i, j, k) = 2^{n_{fw}(i)} \cdot c(i, j, k)$$

Naturally, one can use both "adaptive quantization" techniques (i.e., selective enhancement and frequency weighting) simultaneously (Fig. 3.7). In this case, the values of the resulting coefficients can be expressed as follows:

$$c'(i, j, k) = 2^{n_{se}(k)} \cdot 2^{n_{fw}(i)} \cdot c(i, j, k)$$

In Fig. 3.7, two new bitplanes are generated due to the two AQ tools employed. In this example, the original number of bitplanes is 3, and only four macroblocks are shown. One bitplane is generated due to upshifting (dotted arrows) the upper-left macroblock using Selective Enhancement. The other bitplane is generated (the front dotted) due to frequency-weighting by one-bit-plane shifting of the 4×4 lowest-frequency DCT coefficients (which is applied to all blocks of all macroblocks).

While Selective Enhancement can be employed and controlled on a macroblock-by-macroblock basis, the same Frequency Weighting matrix is applied to all macroblocks in the FGS frames. Selective Enhancement is a relative operation in nature. In other words, if a large number of macroblocks are selected for enhancement, there may not be any perceived improvement in quality. However, selective-enhancement of a large number of macroblocks may be used as a simple tool to de-emphasize an undesired (relatively small) region of the frame. When both Selective Enhancement and Frequency Weighting are used, the upshifting operation does not guarantee that a particular selected-for-enhancement macroblock gets scanned earlier[5] as compared with its original scanning order (i.e., prior to AQ). More details on this can be obtained from [25].

The FGS encoder and decoder with the AQ tools described above are illustrated in Figs. 3.3 and 3.4, respectively. At the encoder side, bitplane shifting due to selective-enhancement and frequency-weighting are performed on the residual FGS signal prior to the scanning and entropy coding of the bitplanes. Bitplane deshifting is performed at the decoder side after the entropy decoding process and prior to the computation of the inverse DCT of the FGS residual signal. Finally, we include some visual comparisons to highlight the benefits of AQ in Figs. 3.8 and 3.9.

3.3 HYBRID TEMPORAL-SNR SCALABILITY WITH AN ALL-FGS STRUCTURE

Temporal scalability is an important tool for enhancing the motion smoothness of compressed video. Typically, a base-layer stream coded with a frame rate f_{BL} is enhanced by another layer consisting of video frames that do not coincide (temporally) with the base layer frames. Therefore, if the enhancement layer has a frame rate of f_{EL}, then the total frame of both the base- and enhancement-layer streams is $f_{BL} + f_{EL}$.

In the SNR FGS scalability structure described in Section 3.2, the frame rate of the transmitted video is *locked* to the frame rate of the base layer, regardless of the available bandwidth and corresponding transmission bitrate. Since one of the design objectives of FGS is to cover a relatively wide range of bandwidth variation over IP

---

[5]To be precise, the scanning order of a macroblock is defined here by its nonzero most-significant-bitplane (MSB). Therefore, although all macroblocks are actually scanned in the same order from one bitplane to another, the times at which the macroblocks' MSBs get scanned and coded differ from one macroblock to another.

FIGURE 3.8: Impact of FGS AQ selective enhancement: (left) without AQ and (right) with AQ.

FIGURE 3.9: Impact of frequency weighting: (left) without frequency weighting and (right) frequency weighting used to emphasize low DCT frequencies.

networks (e.g., 100 kbps to 1 Mbps), it is quite desirable that the SNR enhancement tool of FGS be complemented with a temporal scalability tool. It is also desirable to develop a framework that provides the flexibility of choosing between temporal scalability (better motion smoothness) and SNR scalability (higher quality) at transmission time. This, for example, can be used in response to user preferences and/or real-time bandwidth variations at transmission time. For typical streaming applications, both of these elements are not known at the time of encoding the content.

Consequently, the MPEG-4 framework for supporting hybrid temporal-SNR scalabilities building upon the SNR FGS structure is described in detail in [20]. The proposed framework provides a new level of abstraction between the encoding and transmission processes by supporting *both* SNR and temporal scalabilities through a *single* enhancement-layer. Figure 3.10 shows the proposed hybrid scalability structure. In addition to the standard SNR FGS frames, this hybrid structure includes

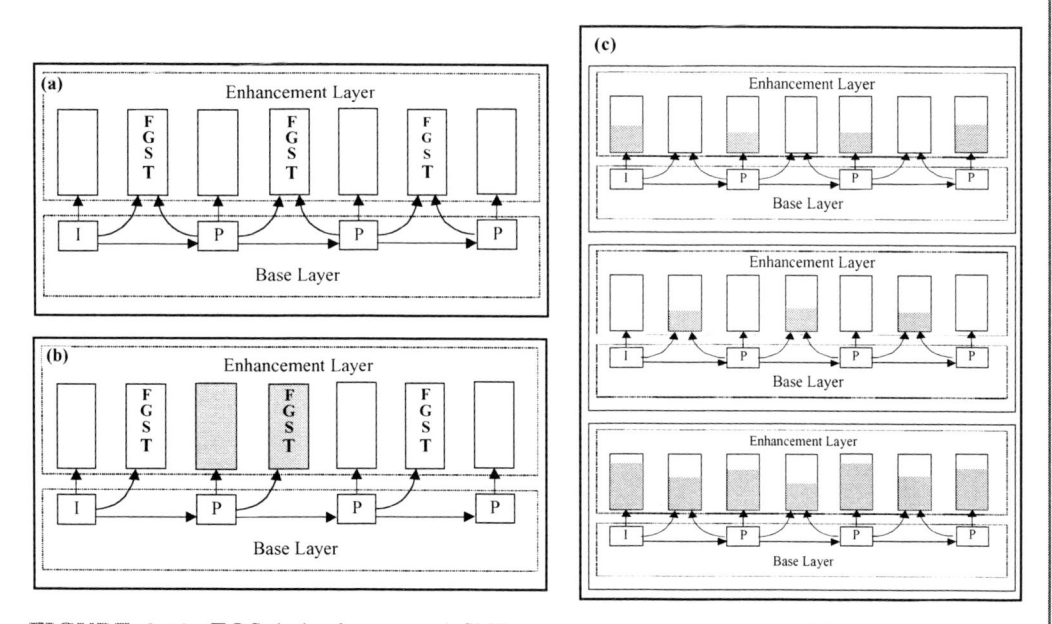

FIGURE 3.10: FGS hybrid temporal-SNR scalability structure with (a) bidirectional and (b) forward-prediction FGST pictures, (c) examples of SNR-only (top), temporal-only (middle), or both temporal-and-SNR (bottom) scalability.

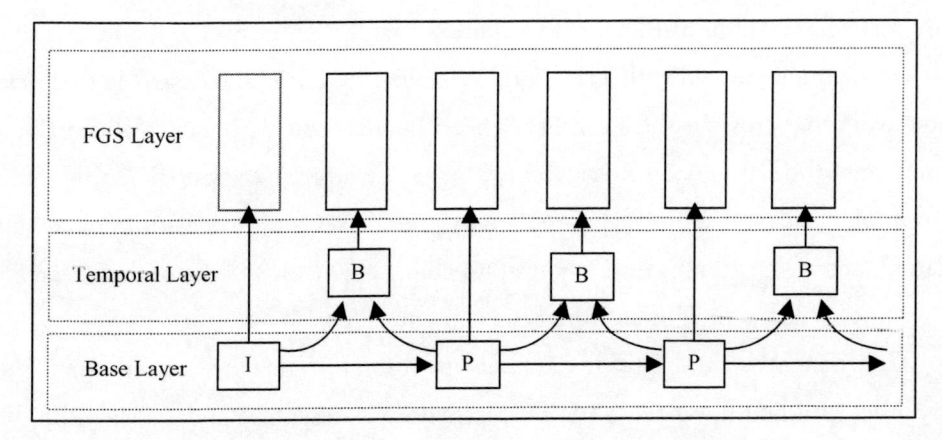

FIGURE 3.11: Multilayer FGS-temporal scalability structure.

motion-compensated residual frames at the enhancement layer. These motion-compensated frames are referred to as FGS Temporal (FGST) pictures [20].

As shown in the figure, each FGST picture is predicted from base-layer frames that do not coincide temporally with that FGST picture, and therefore, this leads to the desired temporal scalability feature. Moreover, the FGST residual signal is coded using the same fine-granular video coding method employed for compressing the standard SNR FGS frames.

Each FGST picture includes two types of information: (a) motion vectors (MVs), which are computed in reference to temporally adjacent base-layer frames, and (b) coded data representing the bitplanes' DCT signal of the motion-compensated FGST residual. The MVs can be computed using standard macroblock-based matching motion-estimation methods. Therefore, the motion-estimation and compensation functional blocks of the base layer can be used by the enhancement-layer codec.

The FGST picture data is coded and transmitted using a data-partitioning strategy to provide added error resilience. Under this strategy, after the FGST VOP header, all motion-vectors are clustered and transmitted before the residual signal bitplanes. The MV data can be transmitted in designated packets with greater protection. More details on hybrid SNR-temporal FGS can be obtained from [20]. Finally, these scalabilities can be further combined in a multilayer manner, and an example of this is shown in Fig. 3.11.

CHAPTER 4

MPEG-4 Video Error Resilience

MPEG-4 was designed to support universal access, and therefore several provisions were made to make MPEG-4 content robust and resilient, so that it could be transmitted to a set of heterogeneous devices over heterogeneous networks with varying loss characteristics. In this section, we focus on the network transmission of MPEG-4 video and describe the particular error resilient features within MPEG-4. Finally, we also briefly mention approaches for packetizing MPEG-4 content, as the payload of Real-time Transport Protocol (RTP) [61] network packets, and some experimental results to highlight the performance of these error resilience tools under lossy channel conditions.

4.1 INTRODUCTION

Video streaming is becoming increasingly popular for a large variety of applications. Whereas previous video coding standards such as MPEG-1 [26], MPEG-2 [27], or H.261 [28] were mainly designed for a specific application and a limited set of transport networks, MPEG-4 targets universal access. Therefore, on the one hand specific tools for existing networks are required; on the other hand the standard should also provide generic features which allow using MPEG-4 in emerging and future transport and network environments without the necessity to add additional versions and releases.

For many applications the distribution of video is accomplished using transport mechanisms of public or private telecommunications networks. Digital networks usually provide only a limited bit rate, and the transmitted data can be lost, altered, or delayed. For some applications it is possible to use network protocols with automatic retransmission of lost or altered data if the time constraints permit. For example, download-and-play applications such as access to digital libraries or video-on-demand are not time-critical and usually the network can guarantee a successful delivery.

However, in real-time scenarios the possibilities of retransmission requests are obviously limited or even impossible. Hence, error resilience in video is necessary especially for multimedia transmission over wired and wireless networks across the Internet as well as wireless tin enterprise networks. We discuss concepts and MPEG-4 features for video transmission in error-prone environments and provide some performance results for real-time video transmission over the Internet as well as wireless networks.

4.2 MPEG-4 VIDEO TRANSMISSION IN ERROR-PRONE ENVIRONMENT

4.2.1 Overview

In the following we restrict ourselves to backward predictive coding using P-VOPs and rectangular frames. More details on error resilience in combination with shape coding may be obtained from [29]. Figure 4.1 presents a simplified yet typical system when motion-compensated predicted video such as MPEG-4 is transmitted over error-prone

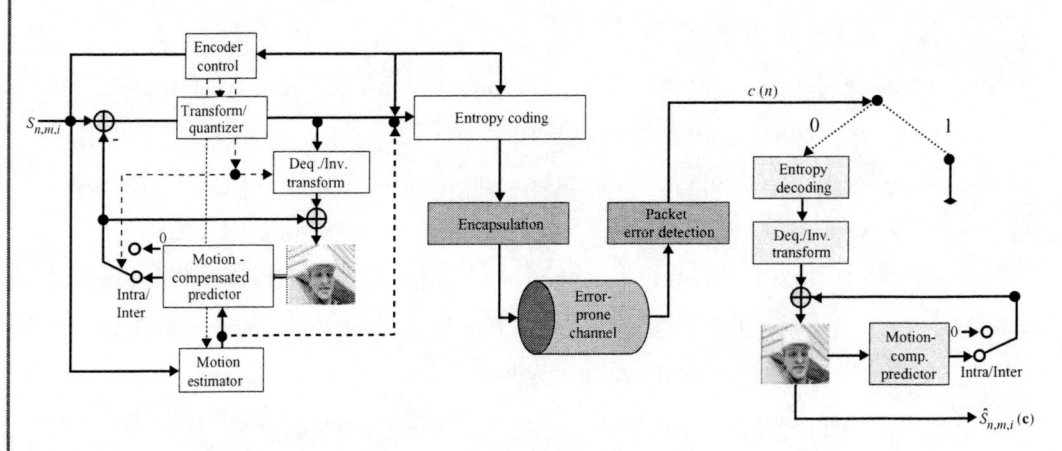

FIGURE 4.1: Motion-compensated video transmission in error-prone environments.

channels. The video encoding process is commonly based on a sequential encoding of frames $n = 1, \ldots, N$. Within each frame, video encoding is based on sequential encoding of macroblocks $m = 1, \ldots, M$.

Assume that all macroblocks of one frame are contained in a single video packet and the video packet is transmitted over the error-prone channel. In addition, let us further assume that corrupted packets are perfectly detected and discarded at the receiver. In case of successful transmission, the video packet is forwarded to the regular decoder operation and the reconstructed frame is forwarded to the display buffer as well as to the reference frame buffer.

When the video packet containing frame n is lost, no decoder operation is specified in the standard. If the decoder just skips the decoding operation then the display buffer is not updated, i.e., the reconstructed frame $n - 1$ is displayed instead of frame n. This will appear as an interruption in the continuous display update. However, as the reference buffer is also not updated, even when frame $n + 1$ is successfully received, this frame will generally not be identical to the reconstructed frame at the encoder. This is because the encoder and the decoder refer to a different reference signal in the motion-compensation process resulting in a so-called *mismatch*. It is obvious that this mismatch propagates to frames $n + 1, n + 2, n + 3$, etc. This phenomenon is called *error propagation*. An example for error propagation is shown in Fig. 4.2: the top row presents the sequence with perfect reconstruction, in the bottom row only frame 21 is lost. Although the remaining frames are again correctly received, the error propagates and is still visible in decoded frame 35.

FIGURE 4.2: Example of error propagation.

The mismatch can be removed only by the successful reception of a nonpredictive I frame. However, I frames lead to added redundancy in the stream. There are additional techniques that can reduce the effects of transmission errors on the video quality. In the following we will introduce the most important principles in error resilient video coding, present some basic methods to realize the features, and discuss their applicability in terms of overhead and delay.

4.2.2 Basic Principles in Error-Prone Video Transmission

4.2.2.1 Avoiding Transmission Errors

The separation principle [30] is a key concept in system design whenever real-time data has to be transmitted over noisy channel. According to this principle, any source signal can be transmitted optimally via a noisy channel even when the source coding is adapted only to the source and the channel coding is adapted only to the channel. Recently, the invention and deployment of powerful channel coding schemes in wireless systems, e.g., Turbo codes [31] and Low-Density Parity Check (LDPC) [32] codes, allow operating very close to the capacity of mobile as well as packet-lossy channels. Error-free transmission with high throughput can also be achieved by applying retransmissions of lost packets. However, these error correction techniques usually add significant delay to the system due to necessary long block length of forward error correction schemes or the retransmission delay. If the real-time constraints of the video do not permit these long delays, packet losses or bit errors cannot be avoided. Then, the operations commonly performed at the video decoder include error detection, resynchronization, concealment, and recovery, and are indicated in Fig. 4.3.

4.2.2.2 Error Detection

The presentation of bit errors can lead to significant performance degradations due to variable length coding, syntax dependencies, as well as spatial and temporal prediction used. Therefore, it is vital to at least detect the errors and then to invoke an appropriate error handling for the corrupt data. Almost all underlying transport protocols provide some kind of error-detection capabilities for at least a syntactical transport unit, which

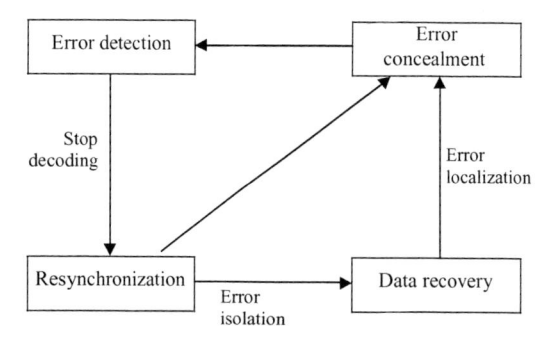

FIGURE 4.3: Advanced decoder operations in case of the reception of erroneous data.

in our case up-to-now includes an entire video frame. Simple receivers use this error indication and drop the entire packet resulting in the effects as shown in Fig. 4.2.

4.2.2.3 Resynchronization

Once a bit error is detected, the decoding can be continued from the next synchronization point which, in general, coincides with the start of a new video frame. However, in many cases bit errors are not spread over the entire coded representation of the frame. Still, if each frame is transmitted within a single syntactical entity, the variable length coding on syntax level does, in general, not allow successful decoding of the remaining data, even if it is error-free. Hence, it is useful to allow resynchronization earlier than at the next coded frame. This can be accomplished by providing additional resynchronization points within the bitstream of a coded frame. However, to fall back into lock step with the encoder, it is not sufficient to just provide a starting point of a variable length code word. Due to semantic dependencies in the decoding process, basically all information in the frame header as well as additional information, e.g., the spatial position, is necessary. Finally, complete semantic resynchronization within one frame also requires that prediction over macroblock boundaries within one frame is not used. Obviously, these methods have associated overheads in terms of signalling costs and reduced prediction.

4.2.2.4 Data Recovery

With error detection and early resynchronization, most erroneous macroblocks can be identified and some error concealment can be performed. However, in case of a wireless

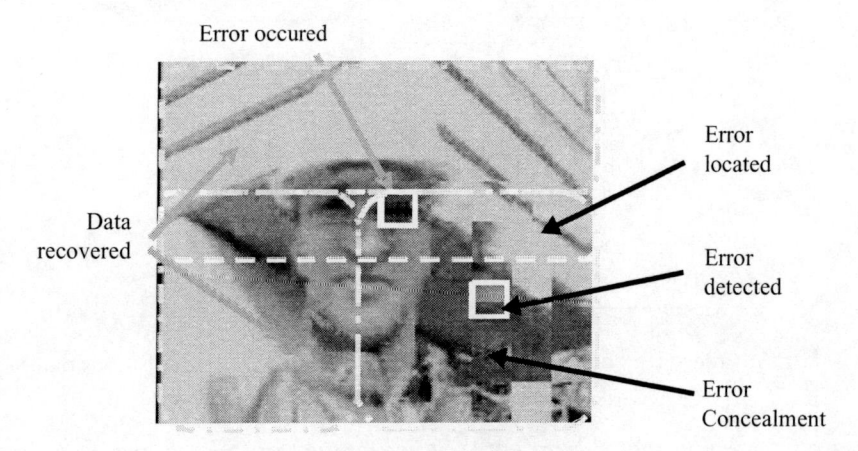

FIGURE 4.4: Decoding corrupted video: error detection, error localization, data recovery, and error concealment.

transmission scenario, where the erroneous bits might be accessible, the corrupted information may still be used in the decoding process when the error can be localized. Besides bit-level detection mechanisms, checks can be performed at other levels to localize errors. For instance, the presence of an invalid VLC table entry, of DCT coefficients and motion vectors that are out of range, of significant visual differences between adjacent macroblocks all indicate the presence of errors. All these do not require additional overhead for error detection; however, they are not always reliable. An example of error detection, localization, and synchronization with data recovery is shown in Fig. 4.4.

4.2.2.5 Error Concealment

For these detected erroneous image parts, an appropriate representation has to be found, e.g., for the lost part in Fig. 4.4. *Error concealment* is a nonnormative feature in any video coding standard, and there exists a vast amount of literature dealing with different levels of error concealment, from very simple to very complex. Typical error concealment schemes use information from the surrounding areas in the spatial and temporal domain to conceal any missing data.

Spatial error concealment methods [33] interpolate from pixel values of spatially adjacent macroblocks [34] to conceal missing data. More advanced methods in the

| Previous frame concealment | Spatial error concealment | Temporal error concealment | Adaptive error concealment |

FIGURE 4.5: Performance of different error concealment strategies: previous frame concealment, spatial concealment [34], temporal concealment [36], and adaptive spatial and temporal concealment.

frequency domain, such as projection onto convex sets (POCS) [35], have also been proposed. Temporal concealment methods use surrounding motion vectors as well as previous frames to find an appropriate reconstruction [36, 37]. The macroblock mode information of the reliable neighbours can be used to decide whether spatial or temporal error concealment or a combination provides the best results. Figure 4.5 shows the performance of different error concealment algorithms.

Previous frame concealment replaces the missing information by the information at the same location in the temporally preceding frame. Clearly, adaptive temporal-spatial error concealment exhibits the least artifacts. In the remainder of this lecture we will use either simple previous frame concealment or the adaptive temporal and spatial error concealment.

4.2.2.6 Reduced Error Propagation

These presented techniques typically cannot eliminate the mismatch in the prediction signal between the encoder and the decoder. Hence, spatiotemporal error propagation can still lead to severe degradations. Although the mismatch decays over time to some extent, the leakage in standardized video decoders is not very strong. Therefore, the decoder has to be provided with other means to reduce or stop the error propagation. The straightforward approach of inserting I frames is quite common for broadcast and streaming applications as these frames are also necessary to randomly access the video sequences. More subtle methods have been considered in different standards. First, intrainformation can be introduced for just parts of a predictively coded image, e.g., a single macroblock is coded in intramode. Second, instead of synchronizing the decoder reference frame to the encoder, the encoder reference frame might be synchronized to the decoder's reference frame by using some kind of feedback.

4.2.2.7 Data Partitioning and Prioritization

Data Partitioning describes all possible methods of dividing the dependent video data into transport units. This includes, among others, spatial subdivision of frames into groups of macroblocks, packetization of I-, P-, and B-frames into transport units (temporal scalability), SNR scalable coding techniques where each layer is packetized in a single packet, and grouping of video syntax elements into partitions. Especially for the latter three cases, the resulting video packets are of different importance for the reconstructed video quality. In many cases the information in some transport units is even useless if some other transport units cannot be decoded. A formalized framework for these coding dependencies has been introduced using dependency graphs [38]. For example, I- and P-frames are more important than B-frames as the latter are not used as reference in the motion compensation process. Similarly, for the SNR scalable coding methods the base layer is more important than the enhancement layer(s). If the network can provide means to prioritize the transmission of important data packets, i.e., to reduce the error probability of these packets, then the division of video data in important and less important information can provide significant benefits. Prioritization in networks can be accomplished by many different means. Examples include, but are not limited to, packet size reduction for more important data, retransmission of important packets, unequal error protection in wireless networks, unequal erasure or loss protection for packet loss channels, adaptation of delivery deadlines for more important data, differentiated services, etc.

However, for low-delay real-time applications, temporal scalability cannot be used, and SNR scalable methods may suffer from reduced coding efficiency as well as complex encoding and decoding algorithms. Therefore, simple but efficient data partitioning methods are desirable in error-prone environments and have been introduced in different video coding standards.

4.3 ERROR RESILIENCE TOOLS IN MPEG-4
4.3.1 Introduction

We now present the error resilience tools introduced in MPEG-4. It is important to understand that the normative part of any video coding standard consists of only the appropriate definition of the order and semantics of the syntax elements and the decoding of error-free bitstreams. Therefore, many features introduced in the standard are not error resilient per se, but require appropriate mechanisms of the transport protocol or the decoder to be fully exploited. These actions are either not standardized at all, or, their specification is outside the video standard, e.g., in a transport protocol specification. If the encoder applies error resilience features in the encoding process, it should be aware that the decoder can at least partly exploit the additional overhead in the bitstream. The appropriate selection of error resilience tools also requires some knowledge on the expected transmission conditions. First, not all features are useful to fight against any kind of transmission errors, e.g., some features suitable for bit-error–prone channels do not provide any benefits for channels with pure packet losses. Second, the knowledge of the characteristics of the channel, e.g., the loss probability or the bit error probability, can help the encoder to select methods appropriately.

4.3.2 Resynchronization and Header Extension Code

The MPEG-4 standard provides means to insert spatially distinct resynchronization points within the video data of a single frame. This allows partitioning the frame into logical partitions containing a sequence of macroblocks. Whereas previous standards such as H.261 and H.263 version 1 permit partitioning only the image into rows of macroblocks, called groups of blocks (GOBs), in MPEG-4 any integer number of macroblocks can form such a logical partition, called video packets (VP). The bitstream syntax when using these video packets is shown in Fig. 4.6.

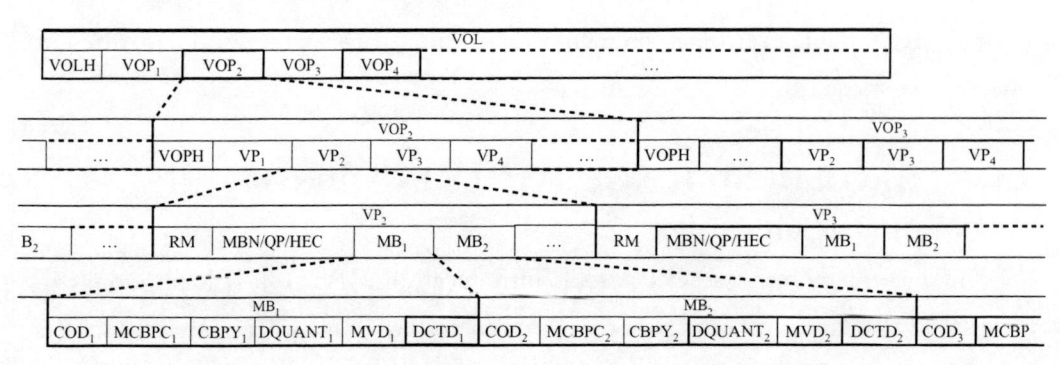

FIGURE 4.6: MPEG-4 video stream with resynchronization markers.

The start of a logical partition is identified by a unique synchronization code word of 17 bits, the *resynchronization marker*. Therefore, if the decoder detects this unique synchronization word, it has automatically detected the start of a new macroblock. However, this is not sufficient to continue decoding at the semantic level. Two additional fields are added after the resynchronization marker. First the absolute macroblock number of the data providing the spatial location of the macroblock in the current image. Second, the quantization parameter for the macroblock is coded using three to nine bits, depending on the quantizer precision.[1] Important header data such as spatial resolution, timing information, and VOP type may also be repeated at this resynchronization point, and its presence is signalled using the header extension code (HEC). Note that the start of new VOP, indicated by the VOP start code and followed by the VOP header, also serves as a resynchronization point.

The introduction of more frequent synchronization possibilities provides the advantages that early resynchronization is possible and errors can be localized as the correct detection of a marker limits the error to the area before the resynchronization point. In addition, if the error can be isolated within a limited amount of macroblocks, the surrounding area below and above the error is likely to still be correct. Then, the information, e.g., motion vectors, of the surrounding correct macroblocks can be used for advanced

---

[1] In addition, to avoid spatial error propagation, the predictive coding of motion vectors is disabled except for the macroblocks between two resynchronization markers.

error concealment techniques. Finally, it is common to encapsulate the information in between two resynchronization markers into a single transport packet.

The encoder can theoretically insert resynchronization points in the video bitstream between any two macroblocks. The resulting video packet can span several rows of a macroblock and can even include partial rows of macroblocks. In general, error resilience improves with the number of resynchronization points; however, these resynchronization markers have associated overheads. The overhead has different reasons: The additional overhead for the resynchronization markers including macroblock number, quantization parameter, and HEC as well as the abandoning of motion vector prediction adversely affect coding efficiency for the video bitstream. In addition, transport protocols use resynchronization markers as natural breakpoints for the start of transport packets. However, as packets need additional bits for packet headers, shorter packets also increase the signalling overhead on transport layers. Therefore, as for almost any error resilience tool, the usage and frequency of resynchronization markers have to be selected carefully to provide optimal quality for a given total bit rate and specific transmission conditions.

Even if markers are inserted at regular macroblock intervals, their actual locations in the bitstream are determined by the variable length coding of the underlying macroblocks. Hence in high-activity regions (where each macroblock requires a large number of bits), these markers are likely to be farther apart, while in low-activity regions, they are likely to be closer. Therefore, the error probability is higher for these high-activity areas, which are generally also harder to conceal, whereas low-complexity areas are lost less likely though being concealed quite easily. The positioning of resynchronization points in the image as well as the resynchronization markers in the bitstream is shown in Fig. 4.7(a).

On the basis of these observations, a different resynchronization mode has been identified to provide significant benefits. In this mode, resynchronization markers are inserted periodically with a distance of approximately K bytes. The encoder inserts a resynchronization marker whenever at least K bytes have been generated starting from the last resynchronization point, and a macroblock boundary is reached. Note that the number K can be chosen arbitrarily, and it is, in general, not known to the decoder. This

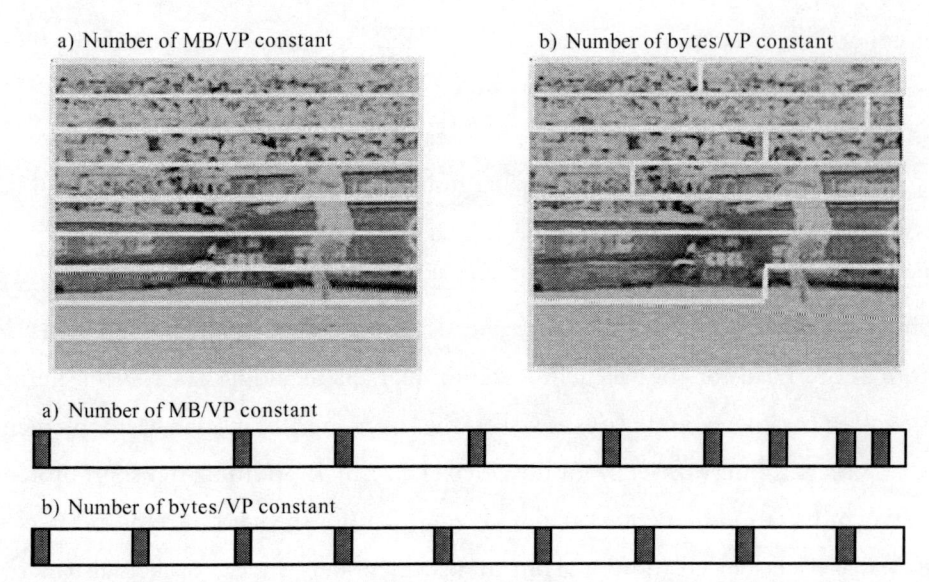

a) Number of MB/VP constant b) Number of bytes/VP constant

a) Number of MB/VP constant

b) Number of bytes/VP constant

FIGURE 4.7: Different resynchronization schemes: (a) the number of macroblocks per video packet is constant resulting in video packets of different length and (b) the number of bytes per video packet is almost constant resulting in different number of macroblocks per video packet.

leads video packets with similar lengths, and therefore, the same probability to be hit by a bit error. Instead, the number of macroblocks within one video packet is arbitrary depending on the complexity and the activity in the area to be encoded. Positions of resynchronization points in image and bitstream for this mode are shown in Fig. 4.7(b).

4.3.3 Data Partitioning

In P-VOPs, the decoding of the DCT coefficients is useful only if the motion vectors have been decoded successfully. In contrast, the motion vectors alone can still be useful as motion compensation can be performed, with the residues set to zero. In the conventional bitstream syntax (see Fig. 4.8), motion vector data and DCT data for each macroblock are coded together, and, in case of error detection all data in the video packet usually has to be discarded.

The data partitioning scheme in MPEG-4 addresses this problem by separating the motion data and the DCT data by a so-called *Motion Boundary Marker* (MBM).

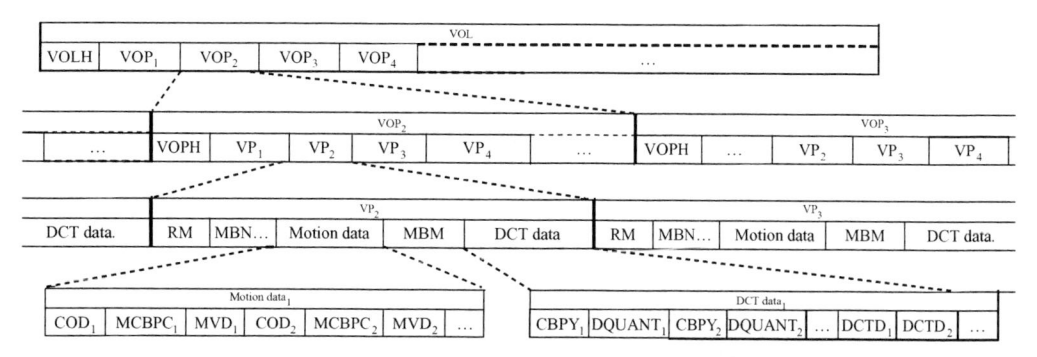

FIGURE 4.8: MPEG-4 video stream with resynchronization markers and data partitioning.

The MBM is a unique 17-bit marker, which does not emulate a valid motion vector. The motion vectors and the macroblock modes such as COD and MCBPC are coded together before the MBM. The information related only to the DCT coefficients, namely the incremental modification of quantization parameter, DQUANT, and the coded block pattern for the luminance component, CBPY, are included in the DCT data part together with the DCT coefficients. For increased efficiency, the motion data of all macroblocks included in a video packet are collected before the MBM, and the DCT data of all macroblocks are collected after the MBM. The exact syntax and order are shown in Fig. 4.8.

As all data elements are coded exactly the same way as in case of a regular video packet, the only additional overhead when using data partitioning results from the introduction of the MBM. If errors have occurred and detected, then the decoder can react as follows. If errors are detected in the motion data, then the entire video packet is discarded. If errors are detected only in the DCT data, but not in the motion part, then at least the motion data is used for motion compensation at the decoder. In addition, the MBM provides an additional mean to validate the correctness of motion vectors and to localize and isolate errors.

4.3.4 Reversible Variable Length Codes

In MPEG-4 whenever a resynchronization point is found, decoding can be continued as all relevant data are accessible, namely the start of a VLC code word, the start of a

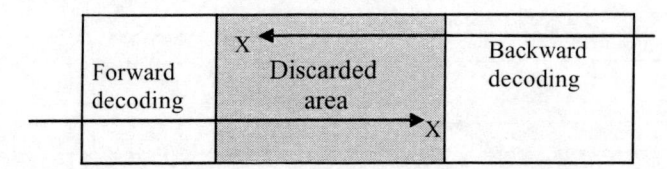

FIGURE 4.9: Discarded data can be reduced by using the forward and reverse decoding.

new macroblock position, etc. Actually, one has also at least some information about the data just before the resynchronization marker, e.g., the end of a VLC code word and the macroblock number. If the decoding could be started in two directions at any resynchronization marker, namely in forward and backward direction, possibly even more data could be recovered. This is shown in Fig. 4.9.

However, this bidirectional decoding is possible only if the variable length code used is not only prefix-free, but also suffix-free, i.e., it is uniquely decodable from both directions. Not all VLC codes fulfil this property, indeed only a small subclass does. MPEG-4 allows the use of reversible variable length codes (RVLCs) for decoding in both directions. In general, RVLCs reduce the coding efficiency; however, the reduction is not too significant as they are designed to match the data statistics. More background and information on the design of RVLCs can be obtained from [39].

Note that for the successful application of backward decoding, it is not sufficient to just provide RVLC codes, but also the semantics have to be such that they can be interpreted when decoded backwards. For instance, care needs to be taken while decoding differentially coded motion vectors or differential quantization information. Their syntax and semantics should be changed to make them reversibly decodable although this comes at the expense of significantly reduced compression efficiency. Of course, the DCT coefficients by themselves are reversibly decodable. The RVLCs are typically combined with the data partitioning mode, where the DCT data are written at the very end of the video packet (see Fig. 4.8). The reversible decoding of RVLCs is not specified in the standard, but several different strategies have been suggested on how to use the RVLCs in the decoding process for error localization and isolation in case of undetected bit errors. The interested reader is referred to [29].

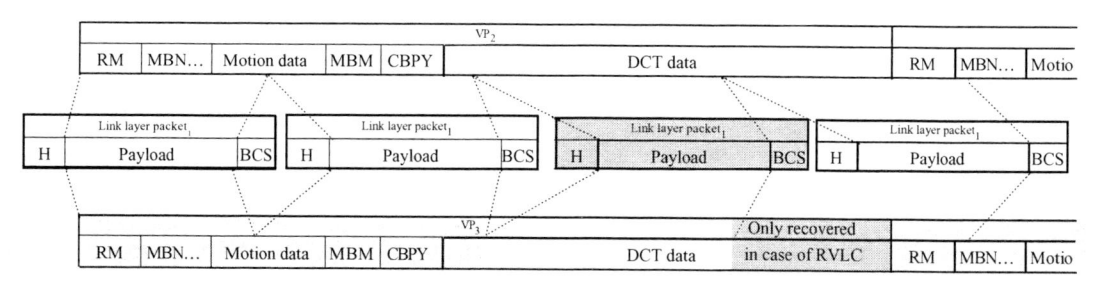

FIGURE 4.10: Transmission of a segmented MPEG-4 bitstream: an example how additionally data can be recovered using data partitioning and RVLCs.

An interesting practical scenario for typical mobile transmission systems is illustrated in Fig. 4.10. If the video packet is segmented into smaller link layer packets, each with a separate block check sequence (BCS) to check for errors in the small packet, then additional reliable means to localize the errors are provided. In this example, with loss of one video packet, the later part of the video packet can be recovered only with the application of the RVLC.

4.3.5 Intrarefresh

MPEG-4 allows inserting I-VOPs in the video sequence at any time. However, completely intracoded frames are usually not favorable in real-time and conversational video applications as the instantaneous bit rate and the resulting delay may be significantly increased. Instead, MPEG-4 permits encoding of single macroblocks in P-VOPs in intramode. This feature was originally introduced for increased compression efficiency as it allows the efficient encoding of uncovered areas and scene changes.

However, this tool can also successfully be used for increased error resilience to limit error propagation. The encoder can insert intramacroblocks appropriately adapted to the video content (to improve coding efficiency), and additionally, possibly adapted to the expected transmission conditions (to improve error resilience).

The insertion of intracoded macroblocks for error resilience can be done either randomly or preferably, using certain update patterns as in [40–42]. The frequency of the updates can be chosen depending on the complexity of the video sequence as well as

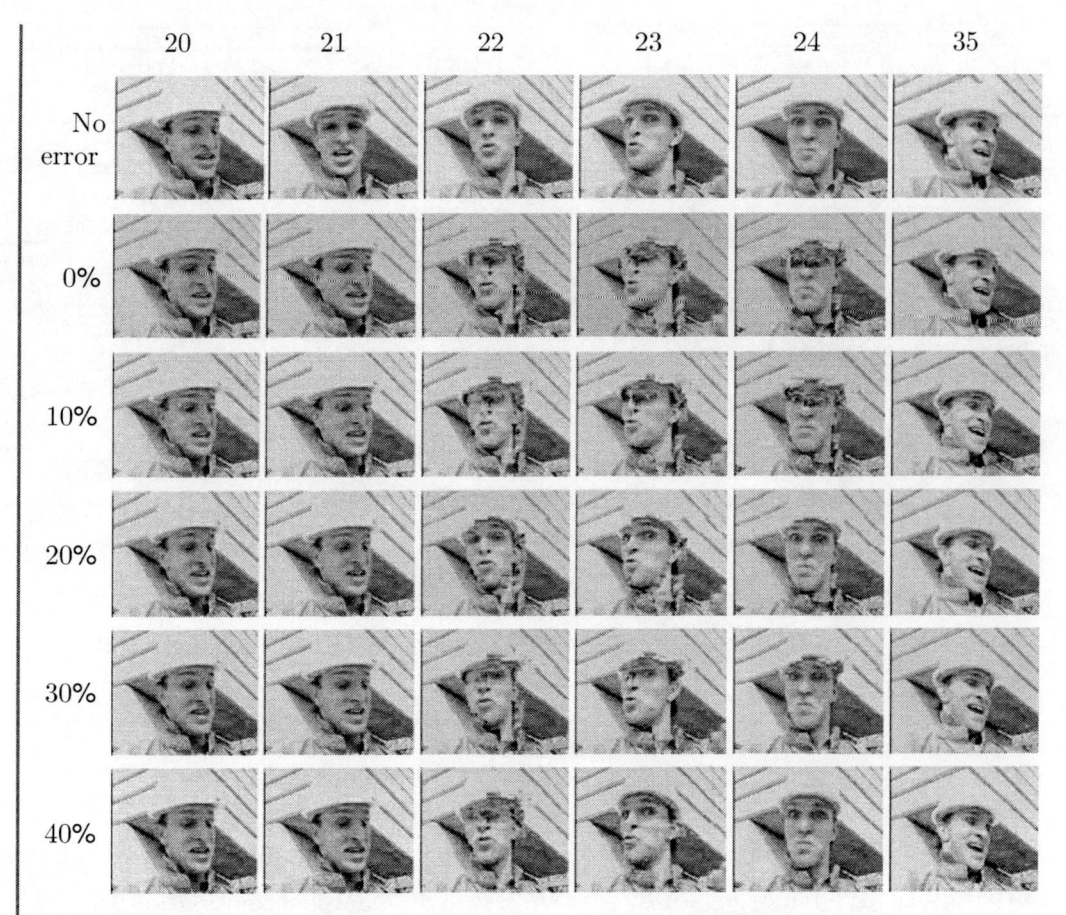

FIGURE 4.11: Coding performance and error propagation for QCIF Foreman sequence is encoded at 64 kbps and 7.5 fps applying different intra refresh ratios.

based on the transmission conditions. More advanced approaches use a rate-distortion optimized mode selection [43, 44] procedure that may be extended by taking into account the influence of the lossy channel: the encoding distortion is replaced with the expected decoder distortion, or, at least an estimate of it. More details on the computation of the expected distortion may be obtained from [45].

In Fig. 4.11, the QCIF Foreman sequence is encoded at 64 kbps and 7.5 fps with different ratios of random intramacroblock refresh updates. We assume that VOP 21 is lost while all other VOPs are successfully received. The first row shows the error-free VOPs. Each remaining row shows the decoded frames with different percentage of the macroblocks intracoded. We can observe that with increasing intra update ratio while the coding efficiency suffers the error propagation is clearly reduced.

4.3.6 New Prediction

In MPEG-4 version 2, an additional tool called NEWPRED (for new prediction) has been introduced to reduce temporal error propagation. This tool is very similar to the H.263 Reference Picture Selection (RPS) Mode (Annex N) and the Slice Structured Mode (Annex K) [46]. NEWPRED provides fast error recovery especially in bidirectional real-time video applications and relies on the following three principles:

- the decoder can send rather quick messages to the encoder, about whether data in the forward link has been received successfully or not;
- online encoding is performed and the encoder can exploit these back-channel messages; and
- the encoder can select other reference frames (which requires additional memory at the encoder and the decoder).

The back channel itself has to be defined in the transport environment and is not entirely in the scope of the MPEG-4 standard. These backchannel messages indicate which NEWPRED (ND) segments (being either an entire VOP or a video packet) have been successfully decoded, and which ND segments have not. When the NEWPRED mode is turned on, the reference used for prediction by the encoder will be updated adaptively according to feedback from the decoder. On the basis of the feedback, the encoder will use either the most recent ND segment or a spatially corresponding but older ND segment for prediction.

Basically the encoder can operate in two different modes when using NEWPRED [47]. These are shown in Fig. 4.12. The left-hand side of Fig. 4.12 shows the ACK mode,

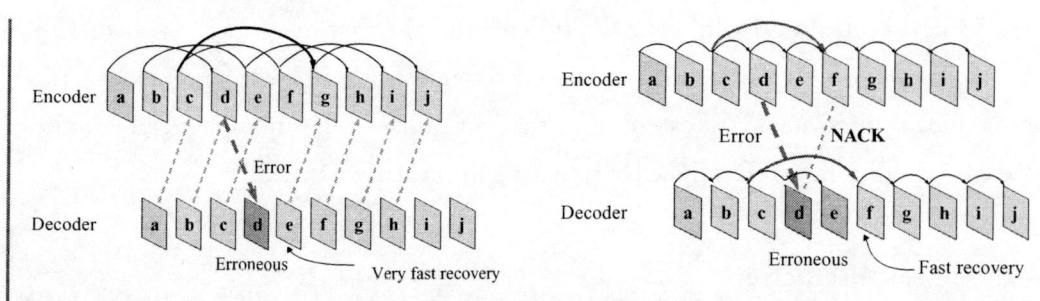

FIGURE 4.12: Operation modes of encoder using NEWPRED: ACK mode (left) and NACK mode (right).

where each correctly received video packet (in this case an entire VOP) is acknowledged. The encoder selects only ND segments for prediction which have been acknowledged. Although in the presentation of a single frame an error might be visible, error propagation and reference frame mismatch between encoder and decoder can be completely avoided, independent of the error concealment applied at the decoder. The efficiency of the mode is low when the round-trip delay is large, as the time distance between the reference area and the area to be coded gets large and the correlation decreases.

For other cases, the NACK mode (presented on the right-hand side of Fig. 4.12) is preferable. In this mode, the encoder alters only its operation if it receives a NACK. Then, the encoder selects ND segments before the ND segment for which the NACK was received. Error propagation occurs over a few frames but is stopped with the successful reception of an ND segment, which references an error-free ND segment. If the feedback delay is too high, the encoder might also decide at times to send an intracoded ND segment. However, as NEWPRED does not necessarily need intrarefresh to stop error propagation, it provides higher coding efficiency although the spatial prediction over ND segments is also limited.

The performance of error resilient video in MPEG-4 version 2 including NEWPRED was evaluated using subjective tests. The application of NEWPRED outperforms simple intraupdates in the recovery time from transmission errors with very low overheads and under low delay conditions.

4.4 STREAMING PROTOCOLS FOR MPEG-4 VIDEO—A BRIEF REVIEW

4.4.1 Networks and Transport Protocols

The two most popular transport networks for media in the past few years are the Internet and wireless networks. As already mentioned, both networks have the inherent problems that the data cannot be transmitted reliably to the receiver, especially if real-time constraints have to be met. We will briefly present the integration of real-time MPEG-4 video into these networks and present some experimental results.

4.4.2 MPEG-4 Video over IP

4.4.2.1 RTP Payload Formats and Packetization Rules

For time-related media delivery over packet-switched networks, the protocol family provides the Real-Time Transport Protocol (RTP). RTP allows transporting, multiplexing, and synchronizing any kind and any mix of media data by using lower layer transport protocols that provide framing. Commonly, UDP on IP is used as transport layer for RTP packets.

RTP provides some common functionalities necessary for all applications. The specific support for different coding types such as (G.72x, MPEG-x, H.26x) is provided by the definition of RTP payload formats. There exist several RTP payload specifications to transport MPEG-4 video over IP. Among these is RFC3016 [48] that specifically targets carrying of MPEG-4 Audio and Visual bit streams without using MPEG-4 Systems.

As MPEG-4 video is basically defined in stream syntax rather than in packet syntax, the bit stream has to be fragmented to map different parts of bit stream to individual RTP packets. RFC3016 provides specifications for the use of RTP header fields and specifies fragmentation rules for mapping MPEG-4 Audio/Visual bit streams onto RTP packets.

4.4.2.2 Selected Performance Results

The video coding expert group (VCEG) of the ITU has adopted a set of common test conditions when transmitting real-time video over the wired Internet [49]. The test conditions also include simplified offline network simulation software, which uses appropriate error patterns captured under realistic transmission conditions. Anchor video sequences, appropriate bit rates, and evaluation criteria are specified. In the following, a representative selection of the common Internet test conditions is used to test the performance of MPEG-4 video. The QCIF sequence Foreman is encoded at a frame rate of 7.5 fps applying only temporally backward referencing motion compensation such that the resulting total bit rate, including a 40-byte IP/UDP/RTP header, matches exactly 64 kbps. The average luminance PSNR (across all frames) is chosen to measure performance. To obtain sufficient statistics at least 15 000 video frames for each experiment were transmitted by looping the encoded sequence and applying a simple packet loss simulator and Internet error patterns.[2] Additionally, it is assumed that the initial I-frame is correctly received.

In the following we present just a very small selection of results that show the performance of different error resilience tools when transmitting over the Internet. An MPEG-4 video codec operating in simple profile was used with previous frame error concealment. The encoder allows inserting resynchronization markers and macroblock intraupdates applying pseudorandom patterns with a certain ratio. Data partitioning and RVLC was not used. The results are not necessarily intended to show the entire potential of MPEG-4, but to provide some insights into the behavior when transmitting video over packet-lossy networks.

Figure 4.13 shows the results for different error rates, different encoder configurations, and different error concealment strategies. For error-free transmission, we can see the best performance is obtained without any error resilience tools. With the introduction of 20% intraupdates the error-free PSNR decreases by about 1 dB. With the introduction of resynchronization markers every 100 bytes the PSNR decreases further due to the introduction of additional IP/UDP/RTP headers. It can be seen that as soon as errors are introduced, intraupdates are very helpful. In contrast, smaller packet sizes (due to

---

[2]The Internet error patterns have been captured from real-world measurements and result in packet loss rates of approximately 3%, 5%, 10%, and 20%. These error probabilities label the packet error rate in Fig. 4.13. Note that the 5% error file is more bursty than the others resulting in somewhat unexpected results.

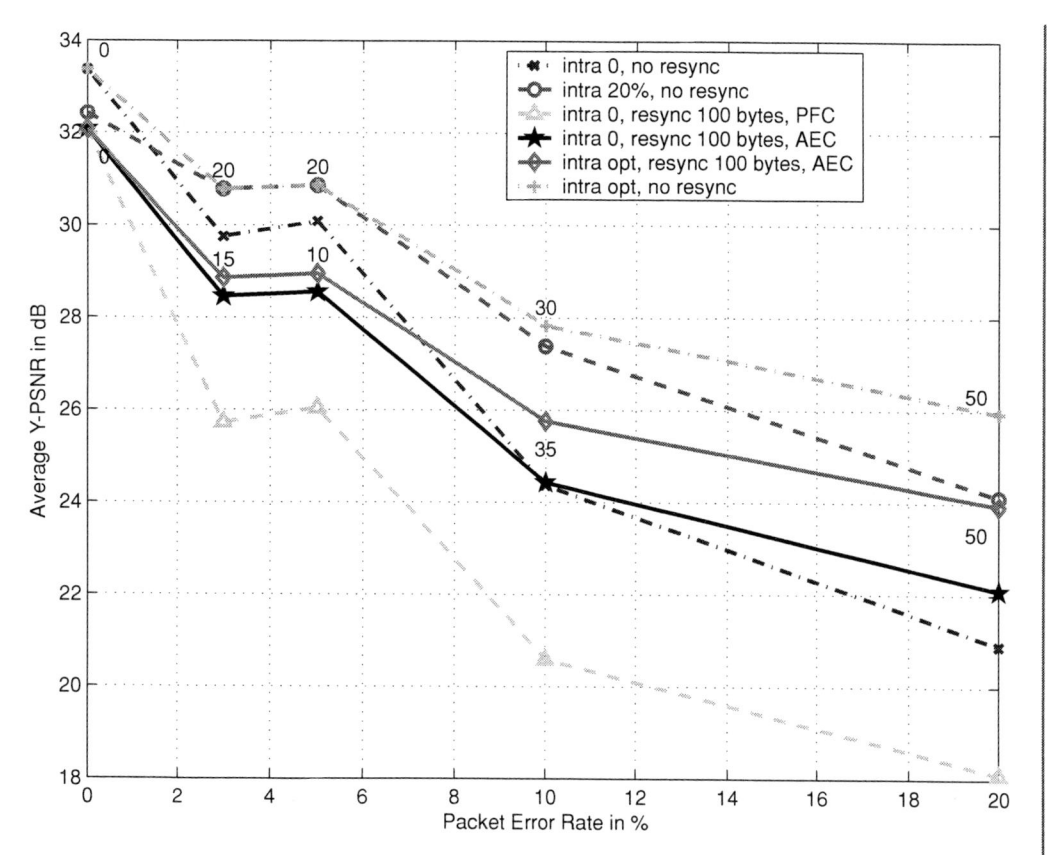

FIGURE 4.13: Results for Internet transmission according to the test conditions in [49] for QCIF test sequence Foreman, 64 kbps, 7.5 fps.

the markers) do not help to improve the performance in pure packet-lossy transmission environments. Furthermore, Advanced Error Concealment (AEC), based on adaptive temporal-spatial concealment, significantly outperforms Previous Frame Concealment (PFC). For this test sequence, we also use channel adaptive intraupdate ratios, represented by the Intra Opt curves. We vary the intra update ratios based on the loss rate, and the update ratios are indicated on the curves. We observe that a 20% intra update ratio is well suited for 3% and 5% error rates, whereas for higher error rates a higher intra update ratio provides better performance as indicated by the numbers above the points.

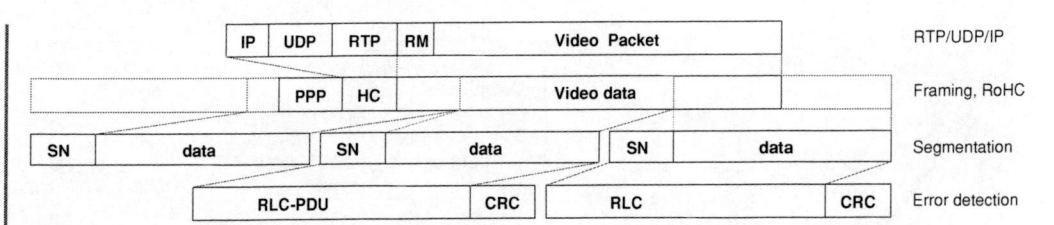

FIGURE 4.14: Packetization of video data for through the 3G user plane protocol stack.

4.4.3 MPEG-4 Video over Wireless

4.4.3.1 Protocol Stack

For packet-switched services, third-generation partnership project (3GPP), responsible for the specification of wireless systems, agreed that the IP-based protocol stack will be used in packet-switched 3G mobile services [50, 51]. Figure 4.14 shows a typical packetization of an MPEG-4 video packet encapsulated in RTP/UDP/IP through the 3G user plane protocol stack.

More details may be obtained from [52].

4.4.3.2 Performance Results

The VCEG of the ITU has refined the MPEG-4 verification test conditions and adopted appropriate common test conditions for 3G mobile transmission of packet-switched conversational and streaming services [53]. The common test conditions define six test case combinations for packet-switched conversational services over 3G mobile networks. In addition, the test conditions include simplified offline 3GPP/3GPP2 simulation software, programming interfaces, and evaluation criteria. Radio channel conditions are simulated with bit-error patterns, which were generated from mobile radio channel

TABLE 4.1: Characteristics of Bit-Error Patterns for UMTS

NO.	BIT RATE	LENGTH	BER	MOBILE SPEED	APPLICATION
3	64 kbps	180 s	5.1e-4	3 km/h	Conversational
4	64 kbps	180 s	1.7e-4	50 km/h	Conversational

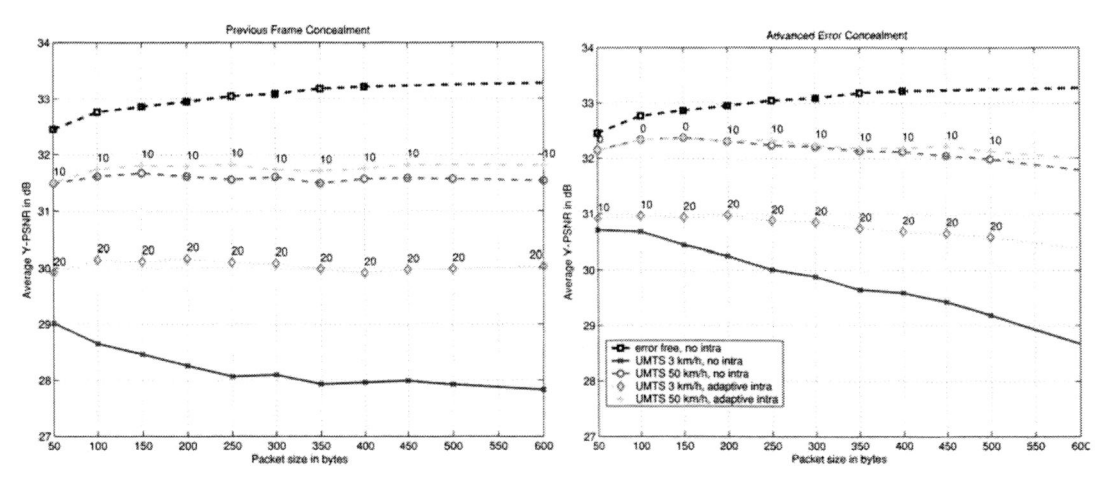

FIGURE 4.15: Results for 3G wireless transmission according to the test conditions in [53] for QCIF test sequence foreman, 64 kbps, 7.5 fps, different mobile speed and different error concealment strategies. Average PSNR is shown over the packet size.

simulations. The properties bit rate, length, bit-error rate, and mobile speed of the bit-error patterns are provided in Table 4.1.

In the following we will present simulation results based on these test conditions for different error resilience and error concealment features. The same video codec is used for the Internet tests. The PSNR is averaged across all frames over 256 transmission and decoding runs. The starting positions for the error patterns, the RTP/UDP/IP overhead after RoHC [62], and the link layer overhead is taken into account in the bit-rate constraints according to [53]. Again, we present results for the QCIF Foreman (300 frames) coded at a constant frame rate of 7.5 Hz.

Figure 4.15 shows the average luminance PSNR over the video packet length in bytes for different speeds of the mobile terminal for PFC as well as for AEC. Rather than explain all the results, we just contrast these results with the Internet case. We observe that the influence of the packetization overhead on the encoding performance is small and less than 0.5 dB in PSNR, unlike for the Internet case. In general, for wireless transmission the introduction of shorter packets (through resynchronization markers) is beneficial. Shorter packets should be supported by advanced error concealment as well as at least some amount of intrarefresh.

CHAPTER 5

MPEG-4 Deployment: Ongoing Efforts

MPEG-4 has encountered significant obstacles on its way to become a de facto industry standard for video coding. These obstacles stem not from the limitations of the technology, but more due to differences about the patent licensing program and the lack of an acceptable compromise between industry leaders. In spite of these differences, MPEG-4 video is supported by all major players in the multimedia arena. Microsoft Windows Media software contains an implementation of the simplest and lowest cost Visual Profile in MPEG-4. Real Networks supports MPEG-4 through a certified plug-in from Envivio. Finally Apple QuickTime (version 6 and beyond) also includes MPEG-4 support.

Unlike in the video coding domain, MPEG-4 is gaining significant deployment in networked multimedia applications. This is due to both the scalability and error resilience support inherent in MPEG-4, and significant industry efforts to develop standard strategies for packetization and streaming of MPEG-4 content. In particular, three separate industry consortia—the Internet Streaming Media Alliance (ISMA), the 3GPP alliance, and MPEG4IP—have focused their efforts on developing open streaming standards. ISMA provides a forum for creating end-to-end specifications that define an interoperable implementation for streaming rich media (video, audio, and associated data) over IP networks. ISMA uses existing standards, contributes to those still in development, produces its own technical specifications when building blocks are missing, and makes those available for the market. The functional areas included in the ISMA 1.0 specification include

- audio format: MPEG-4 High Quality Audio Profile (AAC-LC and Celp);
- video format: MPEG-4 Part-2 Video, Simple and Advanced Simple Profile;
- media storage: MPEG-4 File Format;
- media transport: RTP, RTCP;
- media control and announcement: RTSP, SDP.

The ISMA 2.0 specification, which uses MPEG-4 part 10 for video coding instead of MPEG-4 Part 2, has also been recently released. MPEG-4 part 10, also known as Advanced Video Codec (AVC), was jointly developed by ISO MPEG and ITU, and is currently the most efficient (in terms of compression) video coding standard.

Like the ISMA, the 3GPP [63] also generates specifications for multimedia streaming; however, the 3GPP specifications are targeted toward streaming over 3G wireless networks. The MPEG4IP project was recently created to provide an open-source standards-based system for encoding, streaming, playing, and even broadcasting MPEG-4 encoded audio and video. The project integrates numerous open source applications to provide an end-to-end solution. More information on these different efforts may be obtained from [52, 54–56].

Finally, we would like to list some applications that are driving the deployment of MPEG-4 solutions. These applications range from consumer electronics to business solutions. In the consumer electronics space, the applications include networked DVD players and personal video recorders (PVRs), portable audio players, HDTV displays, smart phones, and digital set-top boxes. In the business space, the key applications include IP-based TV (IPTV) services, video on demand, Internet music stores, and multimedia educational and training applications within enterprise networks.

We believe that the future for deployment of MPEG-4 codecs, file format, and systems is very promising, especially with the prevalence of mixed media interactive streaming applications. Furthermore the technologies developed during the MPEG-4 standardization process have created a fertile field for further research and exploration and led to significant advancements in the field of video coding and streaming, and beyond.

References

1. R. Koenen, "MPEG-4 or why efficiency is more than just a compression ratio," http://www.broadcastpapers.com/sigdis/IBCIntertrustMPEG4Compression Ratio.pdf.
2. MPEG 4: ISO/IEC International Standard, 14 496-2 "Information technology—coding of audio-visual objects: Visual," 1998.
3. R. Koenen, "MPEG-4 Overview", MPEG Output document, N4668, Jeju, March 2002.
4. MPEG Information Web Page http://mpeg.telecomitalialab.com
5. T. Ebrahimi and F. Pereira, "The MPEG-4 Book," Prentice-Hall, 2002.
6. T. Sikora, "The MPEG-4 video standard verification model," *IEEE Trans. Circuits Syst. Video Technol.*, vol. 7, no. 1, February 1997.
7. J. Ostermann, E. Jang, et al., "Coding of arbitrarily shaped video objects in MPEG-4," in *Proc. ICIP 1997*.
8. J. Ostermann, "Coding of arbitrarily shaped objects with binary and grey-scale alpha maps: What can MPEG-4 do for you?", in *Proc. ISCAS 1997*.
9. N. Brady, F. Bossen, and N. Murphy, "Context based arithmetic encoding of 2D shape sequences," in *Special session on shape coding, ICIP 97*, Santa Barbara, 1997.
10. N. Yamaguchi, T. Ida, and T. Watanabe, "A binary shape coding method using Modified MMR," in *Special session on shape coding, ICIP 97*, Santa Barbara, 1997.
11. K. J. O'Connell "Object-adaptive vertex-based shape coding method," *IEEE Trans. Circuits Syst. Video Technol.*, vol. 7, no. 1, February 1997.
12. T. Kaneko and M. Okudaira, "Encoding of arbitrary curves based on the chain code representation," *IEEE Trans. Commun.*, vol. 33, July 1985.
13. S. Lee, D. Cho, et al., "Binary shape coding using 1D distance values from baseline," in *Special session on shape coding, ICIP 97*, Santa Barbara, 1997.

14. T. Chen, C. Swain, and B. Haskell, "Coding of subregions for content based scalable video," *IEEE Trans. Circuits Syst. Video Technol.*, vol. 7, pp. 256–260, February 1997.

15. N. Grammilidis, D. Beletsiotis, and M. Strintzis, "Sprite generation and coding in multi-view image sequences," *IEEE Trans. Circuits Syst. Video Technol.*, vol. 10, no. 2, March 2000.

16. M.-C Lee, W.-G Chen, et al., "A layered video object coding system using sprite and affine motion model," *IEEE Trans. Circuits Syst. Video Technol.*, vol. 7, no. 1, February 1997.

17. A. Smolic, T. Sikora, and J.-R. Ohm, "Long-term global motion estimation and its application for sprite coding, content description, and segmentation," *IEEE Trans. Circuits Syst. Video Technol.*, vol. 9, no. 8, December 1999.

18. P. Selembier and F. Marques, "Region based representations of image and video: Segmentation tools for multimedia services," *IEEE Trans. CSVT*, vol. 9, no. 8, December 1999.

19. H. Radha, M. van der Schaar, and Y. Chen, "The MPEG-4 fine-grained scalable video coding method for multimedia streaming over IP," IEEE *Transact. Multimedia*, vol. 3, no. 1, March 2001.

20. M. van der Schaar and H. Radha, "A hybrid temporal-SNR fine-granular scalability for internet video," *IEEE Trans. Circuits Syst. Video Technol.*, March 2001.

21. W. Li, "Overview of fine granularity scalability in MPEG-4 video standard," *IEEE Trans. Circuits Syst. Video Technol.*, vol. 11, no. 3, pp. 301–317, 2001.

22. F. Ling, W. Li, and H. Sun, "Bitplane coding of DCT coefficients for image and video compression," *SPIE Visual Communications Image Proc.*, vol. 3653, 500–508, 1999.

23. H. Radha, Y. Chen, K. Parthasarathy, and R. Cohen, "Scalable internet video using MPEG-4," *Signal Proc. Image Communication*, vol. 15 (1999) pp. 95–126, September 1999.

24. M. van der Schaar, Y. T. Lin, "Content-based selective enhancement for streaming video," *Proc. IEEE Int. Conf. Image Proc. (ICIP)*, October 2001.

25. S. Peng and M. van der Schaar, "Adaptive frequency weighting for fine granular scalability," in *Proc. Visual Communication and Image Proc. (VCIP)*, January 2002.
26. MPEG-1.
27. MPEG-2: ISO/IEC International Standard 13 818-2, "Generic coding of moving pictures and associated audio information: Visual," 1994.
28. ITU-T Recommendation H.261, "Video codec for audiovisual services at p*64 kbits/sec."
29. R. Talluri, "Error-resilient video coding in the ISO MPEG-4 standard," *IEEE Commun. Mag.*, vol. 36, pp. 112–119, June 1998.
30. C. E. Shannon, The Mathematical Theory of Communication, University of Illinois Press, Urbana, IL, 1948.
31. C. Berrou, A. Glavieux, and P. Thitimajashima, "Near Shannon limit error-correction coding: Turbo Codes," in *Proc. 1993 IEEE Int. Conf. Communications*, Geneva, Switzerland, pp. 1064-1070, May 1993.
32. R. G. Gallager, "Low-Density Parity-Check Codes," *IRE Trans. Info. Theory*, IT-8, pp. 21–28, January 1962.
33. S. Aign and K. Fazel, "Temporal & Spatial Error Concealment Techniques for Hierarchical MPEG-2 Video Coder," in *Proc. of the IEEE ICC'95*, Seattle, pp. 1778–1783.
34. P. Salama, N. B. Shroff, and E. J. Delp, "Error concealment in encoded video," in Image Recovery Techniques for Image Compression Applications, Kluwer Publishers, 1998.
35. H. Sun and W. Kwok, "Concealment of damaged block transform coded images using projections onto convex sets," *IEEE Trans. Image Proc.*, vol. 4, pp. 470–477, April 1995.
36. W. M. Lam, A. R. Reibman, and B. Liu, "Recovery of lost or erroneously received motion vectors," in *Proc. ICASSP*, vol. 5, pp. 417–420, March 1993.
37. M.-J. Chen, L.-G. Chen, and R.-M. Weng, "Error concealment of lost motion vectors with overlapped motion compensation," *IEEE Trans. Circuits Syst. Video Technol.*, vol. 7, no. 3, pp. 560–563, June 1997.

38. P. A. Chou and Z. Miao, "Rate-distortion optimized streaming of packetized media," *IEEE Trans. Multimedia*, vol. February, 2001, submitted.
39. Y. Takishima, M. Wada, and H. Murakami, "Reversible variable length codes," *IEEE Trans. on Comm.*, vol. 43, pp. 158 162, February–April 1995.
40. Q. F. Zhu and L. Kerofsky, "Joint source coding, transport processing, and error concealment for H.323-based packet video," in *Proc. SPIE VCIP*, vol. 3653, pp. 52–62, January 1999.
41. P. Haskell and D. Messerschmitt, "Resynchronization of motion-compensated video affected by ATM cell loss," in *Proc. IEEE ICASSP*, vol. 3, pp. 545–548, 1992.
42. J. Liao and J. Villasenor, "Adaptive intra update for video coding over noisy channels," in *Proc. ICIP*, vol. 3, pp. 763–766, October 1996.
43. T. Wiegand, M. Lightstone, D. Mukherjee, T. G. Campbell, and S. K. Mitra. "Rate-distortion optimized mode selection for very low bit rate video coding and the emerging H.263 standard," *IEEE Trans. Circuits Syst. Video Technol.*, vol. 6, no. 2, pp. 182–190, April 1996.
44. G. J. Sullivan and T. Wiegand, "Rate-distortion optimization for video compression," *IEEE Signal Proc. Mag.*, vol. 15, no. 6, pp. 74–90, November 1998.
45. R. Zhang, S. L. Regunathan, and K. Rose, "Video coding with optimal inter/intra-mode switching for packet loss resilience," *IEEE J. Select. Areas Comm.*, vol. 18, no. 6, pp. 966–976.
46. ITU-T, "Video coding for low bitrate communication," ITU-T Recommendation H.263; version 1, November 1995; version 2, January 1998.
47. S. Fukunaga, T. Nakai, and H. Inoue, "Error resilient video coding by dynamic replacing of reference pictures," in *Proc. IEEE Globecom*, vol. 3, November 1996.
48. Y. Kikuchi, T. Nomura, S. Fukunaga, Y. Matsui, and H. Kimata, "RTP Payload Format for MPEG-4 Audio/Visual Streams," RFC3016, November 2000.
49. S. Wenger, "Common Conditions for wire-line, low delay IP/UDP/RTP packet loss resilient testing," VCEG-N79r1, available from http://standard.pictel.com/ftp/video-site/0109_San/VCEG-N79r1.doc, September 2001.

50. 3GPP Technical Specification 3GPP TR 26.235: "Packet switched conversational multimedia applications; default codecs."
51. 3GPP Technical Specification 3GPP TR 26.937: "Transparent end-to-end packet switched streaming service (PSS); RTP usage model."
52. 3GPP: Third Generation Partnership Project; http://www.3gpp.org.
53. G. Roth, R. Sjöberg, G. Liebl, T. Stockhammer, V. Varsa, and M. Karczewicz, "Common Test Conditions for RTP/IP over 3GPP/3GPP2," ITU-T SG16 Doc. VCEG-N80, Santa Barbara, CA, USA, September 2001.
54. ISMA, ISMA 1.0.1, Internet Streaming Media Alliance Implementation Specification, Version 1.0.1, June 2004.
55. H. Fuchs, N. Färber, "ISMA Interoperability and Conformance," *IEEE Multimedia Magazine*, pp. 96-102, April-June 2005.
56. MPEG4IP: Open Source, Open Standards, Open Streaming; http://mpeg4ip.sourceforge.net/.
57. N. Brady, "MPEG-4 standardized methods for the compression of arbitrarily shaped video objects," *IEEE Trans. Circuits Syst. Video Technol.*, vol. 9, no. 8, December 1999.
58. T. Sikora and B. Makai, "Shape-adaptive DCT for generic coding of video," IEEE Trans. Circuits Syst. Video Technol., vol. 5, no. 1, pp. 59–62, February, 1995.
59. M. Gilge, T. Engelhardt, and R. Mehlan, "Coding of arbitrarily shaped image segments based on a generalized orthogonal transform," *Signal Proc. Image Commun.*, vol. 1, pp. 153–180, October 1989.
60. ITU-T Recommendation H.263, "Video coding for low bit rate communication," 1996.
61. J. van der Meer, D. Mackie, V. Swaminathan, D. Singer, and P. Gentric, "RTP payload format for transport of MPEG-4 elementary streams," RFC3640, November 2003.
62. Hannu, H., Jonsson, L-E., Hakenberg, R., Koren, T., Le, K., Liu, Z., Martensson, A., Miyazaki, A., Svanbro, K., Wiebke, T., Yoshimura, T. and H. Zheng, "Robust

Header Compression (ROHC): Framework and four profiles: RTP, UDP, ESP, and uncompressed," RFC 3095, July 2001.
63. 3GPP Technical Specification 3GPP TR 26.110: "Codec for circuit switched multimedia telephony service; General description."

Biographies

Deepak S. Turaga is currently a research staff member in the Media Delivery Architecture department at IBM T.J. Watson Research Center in Hawthorne, NY. He was previously a Senior Member of Research Staff at Philips Research USA, and a Senior Research Engineer at Sony Electronics. He received a B.Tech. degree in Electrical Engineering from Indian Institute of Technology, Bombay in 1997 and M.S. and Ph.D. degrees in Electrical and Computer Engineering from Carnegie Mellon University, Pittsburgh in 1999 and 2001, respectively. His interests lie primarily in multimedia coding and streaming, and computer vision applications. In these areas he has published over 30 journal and conference papers and one book chapter. He has also filed over fifteen invention disclosures, and has participated actively in MPEG standardization activities. He is a member of several program committees and an active reviewer for different journals and conferences. He is an Associate Editor of the IEEE Transactions on Multimedia.

Mihaela van der Schaar received both the M.S. and Ph.D. degrees from Eindhoven University of Technology, Eindhoven, The Netherlands, in 1996 and 2001, respectively. Prior to joining the UCLA Electrical Engineering Department faculty on July 1st, 2005, she was between 1996 and June 2003 a senior researcher at Philips Research in the Netherlands and USA, where she led a team of researchers working on multimedia coding, processing, networking, and streaming algorithms and architectures. From January to September 2003, she was also an Adjunct Assistant Professor at Columbia University. From July 1st, 2003 until July 1st, 2005, she was an Assistant Professor in the Electrical and Computer Engineering Department at University of California, Davis. Prof. van der Schaar has published extensively on multimedia compression, processing, communications, networking and architectures and holds 22 granted US patents and several more pending. Since 1999, she was an active participant to the ISO Motion Picture Expert Group (MPEG) standard to which she made more than 50 contributions and for which she received two ISO recognition awards. She was also chairing for three years the ad-hoc group on MPEG-21 Scalable Video Coding, and also co-chairing the MPEG ad-hoc group on Multimedia Test-bed. She was a guest editor of the EURASIP Special issue on multimedia over IP and wireless networks and the general chair of Picture Coding Symposium 2004, the oldest conference on image/video coding. She is a senior member of IEEE, and was also elected as a Member of the Technical Committee on Multimedia Signal Processing of the IEEE Signal Processing Society. She was an Associate Editor of IEEE Transactions on Multimedia from 2002-2005 and SPIE Electronic Imaging Journal in 2003. Currently, she is an Associate Editor of IEEE Transactions on Circuits and System for Video Technology and an Associate Editor of IEEE Signal Processing Letters. She received the NSF CAREER Award in December 2004 and the IBM Faculty Award in July 2005. For more information about Prof. van der Schaar and her research, please visit her website at http://www.ee.ucla.edu/~mihaela.

Thomas Stockhammer has been working at the NoMoR Research, Germany, and was visiting researcher at Rensselear Polytechnic Institute (RPI), Troy, NY and at the University of San Diego, California (UCSD) on System and Cross-Layer Design for Wireless Video Transmission and related areas. He has published more than 10 journal and more than 70 conference papers in related areas. He is member of different program committees and regularly reviews papers for different journals and conferences. He also holds several patents and regularly participates and contributes to different international standardization activities, e.g. ISO MPEG, ITU VCEG, JVT, IETF, 3GPP, and DVB and has co-authored more than 100 technical contributions. He is acting chairman of the video adhoc group of 3GPP SA4. In addition, he is also co-founder and CTO of Novel Mobile Radio (NoMoR) Research, a company working on the simulation, emulation, and demonstration of emerging and future mobile multi-user networks and the integration of IP-based applications. Since 2004, he is working as a research and development consultant for Siemens Mobile Devices, now BenQ mobile in Munich, Germany. His research interests include video transmission, cross-layer and system design, forward error correction, content delivery protocols, multimedia broadcast, rate- distortion optimization, information theory, and mobile communications.